Danson House

Danson House

The anatomy of a Georgian villa

Richard Lea and Chris Miele
with Gordon Higgott

Published by English Heritage, The Engine House, Fire Fly Avenue, Swindon SN2 2EH
www.english-heritage.org.uk
English Heritage is the Government's statutory adviser on all aspects of the historic environment.

© English Heritage 2011

The reference numbers for English Heritage images are noted in square brackets in the captions. The front cover, back cover and Figs 0.5, 2.10, 2.11, 4.2, 4.4, 5.5, 6.2, 6.7, 6.10, 6.11, 6.12, 6.17, 6.18. 6.21, 6.23, 6.25, 6.30, 6.34, 6.42, 6.44, 6.46, 6.48, 6.49, 6.50, 6.54, 6.57, 6.58, 9.10, 9.11 and 9.12 are © English Heritage. Fig 0.2 is © English Heritage and appears by permission of Bexley Heritage Trust. Figs 1.1, 1.4, 2.5, 9.1, 9.2 and 9.3 are © English Heritage and appear by permission of Bexley Local Studies and Archive Centre. Figs 4.8, 4.10, 4.11, 4.16, 5.2, 6.31, 6.60, 7.7, 9.6, 9.7 and 9.9 are © English Heritage.NMR. Figs 4.9, 9.4, 9.5 and 9.8 are © English Heritage Photo Library. Figs 3.6, 6.52 and 7.5 are reproduced by permission of English Heritage.NMR. Images under the copyright of other bodies have been credited in their relevant captions.

First published 2011
Reprinted 2012

ISBN 978 1 873592 755
Product code 50763

British Library Cataloguing in Publication data
A CIP catalogue record for this book is available from the British Library.

All rights reserved
No part of this publication may be reproduced or transmitted in any form or by any means, electronic or mechanical, including photocopying, recording, or any information storage or retrieval system, without permission in writing from the publisher.

Application for the reproduction of images should be made to the National Monuments Record. Every effort has been made to trace the copyright holders and we apologise in advance for any unintentional omissions, which we would be pleased to correct in any subsequent edition of this book.

The National Monuments Record is the public archive of English Heritage. For more information, contact NMR Enquiry and Research Services, National Monuments Record Centre, The Engine House, Fire Fly Avenue, Swindon SN2 2EH; telephone (01793) 414600.

Brought to publication by Rachel Howard, Publishing, English Heritage.

Typeset in Charter ITC 9.5pt
Graphics by Richard Lea
Copy edited by Louise Wilson
Indexed by Chris Dance
Page layout by Pauline Hull

Printed in the UK by Butler Tanner and Dennis Ltd.

Front cover: Danson House from the north-east. [DP095878]

Frontispiece: A Sacrifice to Bacchus. The Overmantel painting in the dining room, by Charles Pavillon, 1766. [© Bexley Heritage Trust]

Back cover: The elliptical cantilevered staircase, see Fig 6.57. [DP076965]

CONTENTS

	Acknowledgements	vi
	Preface	vii
1	The origins of Danson	1
2	Augustus and John Boyd	5
3	Robert Taylor, architect of Danson	14
4	Proportion and structure in Taylor's villa at Danson	20
5	Completing the house and landscaping the park, 1765–1773	28
6	Planning, decoration and iconography	33
7	The cost of life as a gentleman	61
8	Nineteenth-century Danson	68
9	Danson House and Park since 1924: decline and restoration	76
	Appendix 1 Transcription of Revd Joseph Spence's comments, May 1763	85
	Appendix 2 Sale Inventory of 1805	88
	Notes	89
	Bibliography	95
	Index	98

ACKNOWLEDGEMENTS

The authors wish to thank all those who helped with their research during the restoration works at Danson House from 1995 to 2002 and the many individuals and institutions who have given advice and practical assistance since then. We are particularly grateful to Dr Gordon Higgott of English Heritage for reordering and editing the whole manuscript and helping to clarify several problems of interpretation.

We owe an especial debt to Ian Jardin and his colleagues Alasdair Glass and John Fidler of English Heritage who directed the project to restore Danson House, and the conservation architects Brian Anderson, Jamie Coath and Dante Vanoli of Purcell Miller Tritton. Helen Hughes, Treve Rosoman, Richard Hewlings and David Robinson at English Heritage contributed important historical and editorial advice as the manuscript evolved into a book text.

The late Commander William Charter, a descendant of the Johnston family, generously placed original documents and drawings at our disposal. Others whose assistance we warmly acknowledge are: Stephen Astley, Celia Fisher, Richard Garnier, Cathy Groves, David Hancock, John Hardy, Ivan Hall, Eileen Harris, Noel Mander, Susan Palmer, R W Sanderson, Mary Schoeser, Michael Snodin, the late Clive Wainwright and Roger White. We thank the Bexley Heritage Trust for its continuous support, encouragement and patience throughout the long gestation of this book, in particular William Roots, Martin Purslow, Janet Hearn-Gillham, Sarah Fosker, Hannah Kay and the staff and volunteers at Danson House. Oliver Wooller, Simon McKeon and the staff of the Bexley Local Studies and Archive Centre have been unfailingly helpful in following up our numerous queries on archival references. Thanks are due to the staff of the Manuscripts Department of the British Library and to the many curators and picture librarians who have helped in the procurement of images, in particular Hugh Alexander, Roxanne Peters, Catriona Cornelius and Justine Sambrook. Their institutions are acknowledged in the captions to the illustrations. In addition, the Sarah Johnston watercolours are courtesy of Commander Charter and family, and the Alfred Bean Portrait and Victorian Danson photographs are courtesy of Mrs Marchant and family.

Special thanks go to Derek Kendall for his superb photography of the building, inside and out, and to June Warrington for coordinating all the illustrations. We are deeply indebted to Rachel Howard in the Publishing Department of English Heritage for seeing the book so expertly into print, aided by Louise Wilson as copy editor, Pauline Hull as designer, and Chris Dance who prepared the index. Finally, we are grateful to all the other English Heritage staff who have helped make this book possible: David Allen, Susie Barson, Clare Broomfield, Duncan Brown, John Cattell, Mathew Dunsdon, Javis Gurr, Cynthia Howell, Irene Peacock, Alyson Rogers, Amanda Rowan, Charles Walker, Lucinda Walker and Nigel Wilkins.

PREFACE

Danson House is an exemplary Palladian villa. Finished to the designs of the architect Robert Taylor in 1766, it was surrounded by an idyllic park of about 81 hectares (roughly 200 acres). The client was John Boyd (1718–1800), a City merchant whose family fortune was founded on West Indian sugar.

Danson was a bold statement that spoke of its owner's social stature and aspirations. Once completed, it became his retreat from the City and a place where he could entertain. Boyd's family and business connections were inseparable, and Danson's location was critical to his success. As the crow flies, Danson is 10½ miles from the City, and the estate borders what was then the principal London to Dover road (Fig 0.1). This easy journey allowed Boyd to enjoy the peace and tranquillity of a country setting while remaining in reach of his lifeline, his import and export business in the City of London.

The country home envisaged by Boyd was very much part of the 18th-century tradition of villa building informed by the celebrated 16th-century architect Andrea Palladio. In his famous treatise, *I Quattro Libri dell'Architettura* [The Four Books of Architecture] (1570), Palladio set out the benefits of the villas he designed for his clients by comparing them with houses in the city.[1] Palladio's villas formed the nuclei of country estates, providing each owner with a country residence and all the ancillary buildings

Figure 0.1
Detail from Andrews, Dury and Herbert, Map of Kent, c 1769, showing the Danson estate lying to the south of the main London to Dover road, and the main house (above the word "Boyd") with wings on each side. [By permission of Bexley Local Studies and Archive Centre]

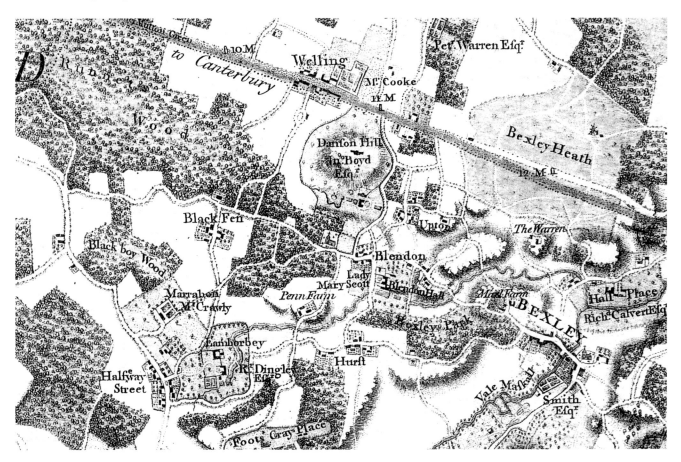

Figure 0.2
George Barrett the Elder. Danson House with the Boyd Family, c 1766, oil on canvas, 1140 x 1680mm. [By permission of Bexley Heritage Trust. (DP094090)]

necessary to the functions of a working farm. Danson did not conform wholly to the Palladian exemplar in this respect, since the farm was built out of sight on the east side of the estate, but in other ways the main building, as originally constructed with attached service wings, followed the Palladian prototype.

Two images record the completed building: a large oil painting by George Barrett Senior (c 1730–84), probably commissioned by Boyd himself in 1766 (Fig 0.2), and an aquatint made from one of Thomas Malton's drawings and published in a suite of engravings of Taylor's work after his death in 1792 (Figs 0.3 and 0.4). In both images the house is flanked by large, single-storey wings, linked to the central villa-block by curving quadrant walls enclosing a forecourt. These wings, however, were not part of Robert Taylor's concept for the building in 1762. As first conceived – and as it appears now – the house is as close to the ideal Georgian villa as any one could imagine – poised, elegant and reserved (Fig 0.5). Extended with the wings, it reflects Boyd's transformation from City merchant to country landowner.[2]

When he began Danson House in 1762, Boyd was an aspiring gentleman in his forties. He had studied theology at Oxford and travelled on the Continent as a young man. He had a keen eye for the arts, choosing Robert Taylor and William Chambers as architects, and Nathaniel Richmond and the critic the Reverend Joseph Spence as landscape designers. In the fitting-out of the house at Danson he revealed a genuine interest in painting, sculpture and music. From his travels in Italy in 1775–6, he brought back one of the greatest archaeological 'finds' of the period, the Piranesi Vase, an 18th-century assemblage of antique marble fragments, nine feet high. In his library, he installed an organ in line with the new fashion set by Handel for musical performances in domestic settings. By the time of his death in 1800, he had amassed a collection of more than 250 paintings, although only a small number of these appear to have been hung at Danson.[3] Once completed, Danson made a remarkably coherent statement, worthy of a gentleman.

As ever, the fortunes of a country house were tied to those of its owner, and Boyd's life in the

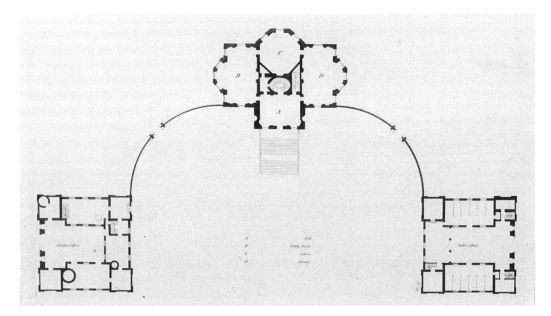

Figure 0.3
Thomas Malton. Engraved plan of the principal floor at Danson, 1792.
[By courtesy of the Trustees of Sir John Soane's Museum. (Malton 12)]

grand manner did not last. The income from his family business, based on West Indian sugar, was severely reduced by the American War of Independence of 1775–83, which prevented sugar from reaching the London markets, and in 1780 Boyd had to mortgage the property to make ends meet.[4] Following Boyd's death, his son, another John, modernised the house, demolishing the wings and building the current stable block between 1802 and 1804, as well as carrying out other works. In 1805, however, his father's debts forced him to sell. The new owner was John Johnston, also a West Indies merchant. In 1863, the Johnstons were succeeded by Alfred Bean, whose wealth was founded on railway building and civil engineering. Both Johnston and Bean died long before their wives, leaving widows ensconced in the grand house. The appearance of the interiors at the point of handover from one family to the other is recorded in a set of watercolours by John Johnston's granddaughter Sarah (*see* Figs 6.13, 6.14, 6.32, 6.56 and 8.6). The Beans again modernised the Georgian interiors, but in a

Figure 0.4
Thomas Malton. Aquatint engraving of Danson, published in 1792, after Taylor's death. The stable and kitchen wings (demolished c 1801–4) were added towards the end of construction, in 1765–6.
[By courtesy of the Trustees of Sir John Soane's Museum. (Malton 13)]

way that proved easier to reverse during recent works of repair and restoration. The descendants of Alfred Bean sold the outlying agricultural land for suburban housing, and many of the streets around Danson Park took their names from the estate's history. However, the park itself, nearly 200 acres (81ha), was kept intact.

The park, house, stable, bridge, lake and the little temple beside it, were sold at auction to Bexley Urban District Council in 1924. Danson then became a popular local attraction like many comparable houses in parks purchased by local authorities. Despite its popularity, the borough did not maintain the building properly, and it gradually degraded until, in 1971, the house had to be closed to the public for reasons of safety. Over the next two decades Danson became a conservation cause célèbre, the most important Georgian 'Building at Risk' in the capital.

In 1995, English Heritage took the property on a 999-year lease from the London Borough of Bexley to oversee and pay for its repair. Initially the project was one of careful conservation, in line with English Heritage's core principles, but this approach changed during the course of investigations on the principal floor, when it became clear that later changes to the fabric of the building were superficial. A great deal of Robert Taylor's original work survived, and where it had been lost to decay or alteration, it was possible to restore the original Georgian character of the building with a high degree of certainty.

Extensive historical and physical investigations were undertaken to inform the re-presentation of the interiors and the programme of external repairs. This book presents the results of this work. The aims of publishing such a study are, first, to mark the successful completion of the repair, and, in the tradition of similar Victorian publications, to commemorate the saving of Danson from ruin. More practically, English Heritage wants to make the results of its technical and historical investigations available to students of Georgian architecture and to practitioners who may themselves be engaged in similar kinds of work.

Figure 0.5
Danson House from the north-east in 2009.
[DP076956]

1

The origins of Danson

Danson enters history as a hamlet, cited in two 14th-century documents as 'Densinton' then 'Danston'. Ridge-and-furrow field patterns and a few building platforms can be observed in the parkland, but not enough to provide a clear image of the medieval landscape. The name, in its various forms, is undoubtedly historic but its origins have not been fully investigated.[1]

By the late 16th century Danson had become a rural estate of about 200 acres (81ha), split between woodland and arable, roughly the size of Danson Park today. In 1571 Matthew Parker, son of the Archbishop of Canterbury, took possession of the estate freehold.[2] Parker's will of 1574 refers to Danson as a 'mannor', though legally it was not. That distinction belonged to Hall Place, nearby. Still, here is evidence of a house at Danson and the first sign of its rising fortunes and its owner's mounting aspirations. In 1761 John Boyd, the man who built the 18th-century house, was more honest, referring to the property as 'the Manor or reputed Manor'.[3]

Some of the best evidence for the pattern of land ownership and the gentrification of early modern and Georgian London remains locked away in antiquarian and topographical writing.[4] However, a pioneering study of Bexley has shown that during the 18th century, the countryside around London received an influx of city dwellers who formed estates of 100 acres or more in an attempt to acquire gentle status. While these holdings tended to be smaller than agricultural estates in the remoter countryside, they changed hands more frequently.[5] Another recent study of Bexley has confirmed these trends.[6] Indeed, the Georgian histories of May Place, Foots Cray Place, Belvedere and Lamorbey follow a trajectory similar to Danson's, inasmuch as each was held for a time by a family whose wealth was derived directly from overseas trade or from the City, rather than from inheritance or land. Even the manor itself, Hall Place, fits this profile, for in 1537 it was purchased by Sir John Champneys, a member of the Skinners' Company, and previously both Sheriff and Lord Mayor of London.

This suburbanisation was matched in other areas of what is now Greater London, for instance, in Kensington, when the Court attracted genteel residences to the area after William and Mary's accession in 1689. Richmond developed slightly later, its popularity assured by the discovery of spa waters in 1696, and the proximity of Hampton Court. Twickenham expanded in the 1720s with speculative terraced housing and a villa for Alexander Pope from 1719, and Horace Walpole's Strawberry Hill from 1748. In such places, ancient houses were rebuilt as villas and others constructed entirely anew. Even distant Epsom was affected by visitors seeking to drink its spa waters. City merchants were in the habit of settling their families there in summer, travelling to town by coach or horse regularly, if not daily, since a return trip took about eight hours. While these western suburbs were densely developed with smaller properties, in the eastern and south-eastern London fringes of Essex and Kent it was easier to amass more extensive holdings. Being less popular, land was cheaper: the exception in the south-east was Greenwich which, like Richmond, was dignified by a royal park.[7]

The history of Danson comes into sharper focus during the late 17th century, soon after the Great Fire of London and at the start of a sustained period of suburban expansion. In 1672 John Adye of Gray's Inn obtained the freehold to the estate, which was still about the size of Parker's a century earlier. Adye appears to have been a property speculator, like many lawyers at the time. His will refers to properties in Covent Garden, around the Strand and near St Paul's Cathedral, as well as estates at Plumstead, Woolwich, East Wickham and Welling.[8] A copy of a lost plan of 1684 shows the so-called 'Mannor of Danson' with boundaries roughly

*Figure 1.1
Estate plan, 1684.
Photograph published in 1925 of an estate plan prepared for the London attorney John Adye, when most of the land was held on leases (the original plan disappeared in 1928). North is at the bottom. The old house at the east (right) end of the canal, is circled. [By permission of Bexley Local Studies and Archive Centre, 728.8DAN. (DP094084)]*

equal to those of the present Park (Fig 1.1).[9] A crucial difference between then and now was the position of the house. It stood, not on the high ridge bisecting the property, as now, but lower down, to the south, near the east (or right) end of a straight ornamental canal, on the site of the present lake. Another plan, dated 1710, shows the entrance to the estate to the east of the canal, at a point midway along the present Danson Road.[10] Years later, a Swedish artist, Elias Martin, made a pen-and-ink sketch of the canal and house – after it had been rebuilt in the late 1740s – as a square-plan, gable-ended structure (Fig 1.2). In 1684, however, canals were newly fashionable garden features in Britain. The earliest known example was constructed in St James's Park in 1660. A far larger canal was begun at Hampton Court in 1661.[11]

John Styleman

In 1697, two years after Adye's death, Francis Styleman, a member of the Joyners' Company, and a London merchant, Mr Rodriguez, took a 200-year lease on the property from Adye's daughter, Mrs Mary Stevens, on behalf of Francis's brother John Styleman (1652–1734),[12] who was then in India. John returned to England in about 1700 and settled at Danson. In 1723, after the death of his third wife, he sublet the house for 99 years to Colonel John Selwyn. Styleman's will, proved in 1734, settled all his property on his fifth wife, including the rental income from Danson of £200 a year. When she died in 1750, this income was vested in a charity set up to build and run 12 almshouses on land near the parish church of St Mary. Styleman's Almshouses, completed in 1755, survive to this day (Fig 1.3).[13]

The repairing lease that Selwyn signed on Danson in 1723 tells us that the property was very well appointed, with more than 250 acres of land including 'ponds, springs, canals, [and] piscaryes'. Selwyn was also obliged to spend at least £1,000 on improving the property within 10 years of taking the lease.[14]

John Selwyn

John Selwyn (1688–1751) was the son of Lt-Gen William Selwyn, Governor of Jamaica. The Selwyn family seat was at Matson, just east of Gloucester (and had been since 1600),

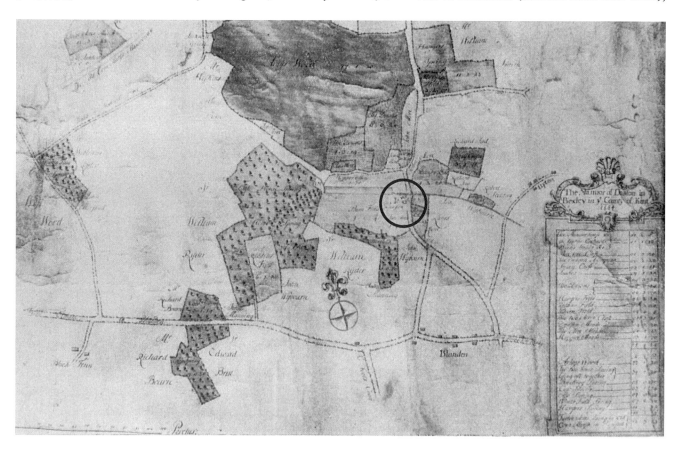

*Figure 1.2
Watercolour by Elias Martin of the park at Danson in about 1768. It shows the canal created by John Adye in the late 17th century, and the house that John Styleman rebuilt in the 1740s.
[Elias Martin: The Park at Danson; NM H 86/1892, © Nationalmuseum, Stockholm]*

and his connections with the area around Danson were through his mother, the daughter of Richard Bettenson of Scadbury, north-east of Chislehurst, and only two miles south of Danson. She was the heiress of her brother, Sir Edward Bettenson, 2nd Bart. Selwyn was married, in around 1709, to Mary, daughter of General Thomas Farrington, who also came from Chislehurst.

Selwyn's early career was in the Foot Guards, in which he was commissioned at birth, rising to be Colonel of the 3rd Foot from 1711 to 1713, while aide-de-camp to the Duke of Marlborough. Through the patronage of the Whig minister Lord Townshend, he obtained a post as Clerk of the Household of the Prince of Wales in 1716, and thereafter both he and his wife were courtiers. He became Groom of the Bedchamber to the Prince in 1718, retaining this post on the prince's succession as king in 1727, until 1730, when he became Treasurer of the Household of Queen Caroline. On her death in 1737 he became Treasurer of the Queen's Pensions and in May 1751 Treasurer to Frederick, Prince of Wales, though only for a few months as he was to die in November of that year. His wife, Mary, was Queen Caroline's Woman of the Bedchamber, reputedly her favourite. Their son, born in 1719 and later to be famous as a wit, was named George Augustus after the Prince. John Selwyn's adherence to Townshend and Walpole, including their period in opposition from 1717 to 1720, was rewarded by the lucrative post of Receiver General and Controller of Customs from 1721 to 1727; and his loyalty to their successor, the Duke of Newcastle, was rewarded by a similarly lucrative post as Paymaster of Marines from 1747 to 1748. In a list published in 1739, his offices were recorded as formerly having been worth £4,600 a year. In 1745 he had a grant from George II of 16 tracts of land in North Carolina, each amounting to 16,000 acres (6,475ha), though exactly what these generated in terms of real income is not clear as the land was unimproved. He was also MP for Gloucester, whose water supply he controlled

*Figure 1.3
Styleman's Almshouses, High Street, Bexley, finished in 1755. This group of 12 almshouses was funded by a charitable trust established in 1750 on the rental income of the Danson estate.
[© Richard Lea]*

Figure 1.4
Lease plan of the Danson estate in 1753, the year John Boyd acquired the property. The plan is aligned with north to the right.
[By permission of Bexley Local Studies and Archive Centre, PEDAN/1/90. (DP094085)]

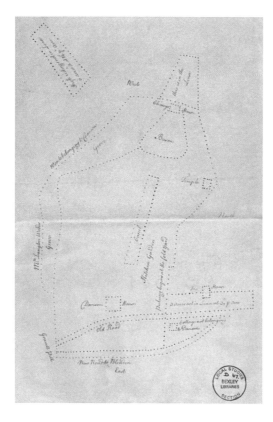

from Matson House, from 1734 to 1751.[15]

Selwyn lived principally in Cleveland Court, by St James's Palace in central London, but conducted much of his parliamentary and other business from Danson.[16] He also spent far more money on improving the estate than he was obliged to by his repairing lease.[17] Between 1745 and his death in 1751, the house was rebuilt on a square plan.[18] It is shown near the bottom of a lease plan made in 1753, next to the 'Old Road', which led north to London, north being on the right side of this plan (Fig 1.4). Elias Martin's sketch of c 1768 shows the house in side view, with its gable end towards the canal (see Fig 1.2), suggesting that the front of the house faced towards this London approach, from where visitors would arrive. An inventory accompanying the lease plan describes, on the ground floor, a hall decorated with maps, a library, a parlour and a drawing room, as well as a bedroom and closet; above were three more bedrooms, a nursery and a 'gallery' with garrets for the servants. The basement included a brick hall with a cellar beneath and two servants' halls, a pantry and a kitchen.[19]

In 1745 and 1746, Selwyn purchased land around the house and moved the road bounding the eastern perimeter (at the bottom of the lease plan) further east again, giving modern-day Danson Road its present alignment (see Fig 1.4; marked 'New Road to Blendon'). A barn and outbuildings shown on an earlier plan had been removed by 1753. Instead the lease plan shows the surviving 'Ice House' north of the house and a square-planned 'Temple' north of the canal. This structure is shown on Joseph Spence's sketch plan of the proposed new park at Danson in 1763 as a domed rotunda, near the 'Great Pond' and reached by a serpentine path from the house at the east end of the canal (see Fig 2.7). The lease plan also shows a 'Bason' at the west end of the canal, and a 'Chineys House' on an island, reached by two bridges. An undated sketch by 'A. C.' shows the 'Chinese Building', and reveals the two bridges to be in the Chinese style as well (Fig 1.5).[20]

Selwyn would have needed an architect for this kind of work, and the person whom he most probably engaged was Thomas Wright (1711–86), a specialist in landscape gardening and ornamental garden buildings in the Chinese style.[21] Wright's diary records that he visited Danson in 1743 and again in 1745, without stating his business there, but it is almost certain that he was providing designs for Selwyn.[22]

Figure 1.5
View of the Chinese Building in the Garden in Danson in Kent, *signed by 'A. C.' The 'Chinese' house and bridges were built for John Selwyn in the 1740s, probably to designs by Thomas Wright. They were swept away by John Boyd 30 years later.*
[The Bodleian Libraries, University of Oxford Gough Maps 13, fol 28, item e]

2

Augustus and John Boyd

John Boyd, the builder of the present Danson House, was the son of a West Indies merchant. A successful merchant himself, he was to use his suburban estate to project an image of refinement far beyond anything his father had aspired to. His ancestry was a mixture of Irish, Scottish (he was distantly related to the 17th-century earls of Kilmarnock) and Huguenot. It was his more recent lineage that shaped him. A branch of the Boyd family had been resident in the Huguenot merchant community in La Rochelle from 1603 when his grandfather, Jean Boyd (1665–1717), left for North America following the revocation of the Edict of Nantes in 1685. In March 1686, Jean, along with his brothers Jacques and Gabriel, arrived in Charleston, South Carolina, where they were among the first settlers.[1]

Jean Boyd's son, Jean Auguste Boyd, later known as Augustus, was born in Carolina in about 1693. By 1700 Augustus had settled on St Kitts, one of the gentle Leeward Isles in the West Indies where his uncle's father-in-law, a French Huguenot, Andrew Thauvet (d 1722), had set himself up as a sugar planter. The Leeward Isles were ideal for sugar cane, one of the 18th century's most valuable crops. Between 1700 and 1770 prices quadrupled, making it, and the rum distilled from it, into one of London's most profitable imports. At first Augustus managed plantations on St Kitts for absentee landlords but by 1718 he had acquired

Figure 2.1
Austin Friars, City of London, c 1881. Watercolour by John Crowther. Augustus and John Boyd set up business here in 1745.
[© City of London, London Metropolitan Archives]

Figure 2.2
Prospect of the slave factory on Bance Island, Sierra Leone, at the mouth of the Sierra Leone River, purchased by Boyd and Co in 1748.
[The National Archives UK, CO 700, Sierra Leone, 1B]

150 acres, cultivated by 49 West African slaves. He also acted as a provisioning agent for planters on other islands, reselling and exporting their produce. In 1718 he married Lucy Peters, the daughter of a wealthy island planter. In the same year, she gave birth to their son John, the future builder of Danson. Apart from this, however things did not go as well for them in St Kitts as they had hoped. First, in 1730, Augustus was unable to take possession of 110 acres left him by Thauvet, and then, in 1732, his bid to gain a seat on the island Assembly failed. Soon after, he, his wife, and their only son decided to leave for London, arriving there in 1735.

John Boyd was 18 years old when he first saw the metropolis of London, and it must have made a great impression on a young man whose whole life had until now been spent on a sugar plantation. London was then by far the largest city in Britain, with approximately half a million people and growing. In Europe, only Paris was remotely comparable, though it has been argued that London's dominance of the nation's life was far greater than that of any other European capital.[2]

The Boyds were lucky to be able to lodge with kin on their arrival – the family of James Pechell, another French Huguenot with business interests in St Kitts. Boyd and Pechell formed a partnership, with offices in Broad Street, then an enclave of Scots, Irish and Huguenot merchants. The pair provisioned colonial planters and acted as their British agents, even reselling or re-exporting their produce, and Boyd himself began to act as a lobbyist, petitioning Parliament on trade bills of interest to himself and his planter clientele. In 1745 he set up on his own in Austin Friars, just south of Broad Street (Fig 2.1). Boyd and Co initially served the same client base but eventually expanded by winning lucrative government contracts to provision the military. This was a bellicose period of British history, first the War of the Austrian succession (1739–48) and then the Seven Years War (1756–63), struggles which left the British economy one of the most powerful in Europe, ensuring its domination of colonial sea trade. From 1747 to 1760, Boyd's company shipped provisions to 14,000 British troops stationed at Kingston in Jamaica. They sent Irish beef and butter through the offices of relations to the increasingly large naval victualling yards at Plymouth and Portsmouth.

The plantations that stood at the head of this trading pyramid were based on slave labour, and Augustus Boyd and his associates became involved, buying a slave factory on Bance Island in the mouth of the Sierra Leone River in 1748 (Fig 2.2).[3] A consortium of West Indies' planters rebuilt the island facilities entirely between 1752 and 1756. It turned a fine profit, peaking in the 1760s when an estimated 10,000 slaves were handled through Bance Island, totalling six per cent of all British slave exports from Sierra Leone. John Boyd remained part-owner of the slave factory until 1784.

We have little information about John Boyd's early life except that in 1737 he went up to Christ Church, Oxford, to read theology, after which he went on the first of several Continental journeys, returning to join his father's business in the family home in Broad Street in the early 1740s. In 1743, John was the first to move out, taking a house in Lewisham High Street, a village then, but one quickly being drawn into the orbit of the capital. His father followed by leasing a much larger property a few doors away, in a prominent, corner position. To judge from 19th-century photographs, this was a large, double-pile, three-storey building (Fig 2.3) called Lewisham House, built in around 1680 for Sir John Lethieullier, Sheriff of the City of London. It was dignified and grand, not a 'villa', but part of a busy village street and old-fashioned by the middle of the 18th century. Although by now comfortably well off, Augustus clearly did not feel the need to display his wealth.[4]

On 27 June 1749, John Boyd married Mary Bumpstead (1734–63), the daughter of a prominent Warwickshire landowner. They were to have five children during Mary's short life: John (1750), Elizabeth (1751), Lucy (1753), Mary (1754), and Augustus (1758), who was to die in his teens. During this time John Boyd began to establish his business interests, serving on the Court of Directors of the East India Company from 1753 to 1764 and as the Company's deputy chairman from 1759 to 1760.[5]

In 1760, Augustus took a lease on the family's first West End house, 4 Great George Street, then newly built and close to the Palaces of Westminster.[6] Boyd and Co's clients had an interest in trade bills, and Great George Street was handy for Parliament. The long terrace of properties on either side of the street had been developed speculatively, with most houses constructed by 1755. The outsides were plain: four storeys with purplish brick frontages and simple stone floor-bands, topped by a stern, classical block cornice. Number 11 on the west side and 29, opposite, on the east, were finished with brick pediments, the former being the only surviving original façade. Rate books show that although Augustus Boyd was the lessee of number 4, neither he nor John ever occupied the property, and that following Augustus's death in 1765 his mother was the owner from 1768 to 1783, even though her permanent residence appears still to have been the family home in Lewisham. In fact, during much of this time, the Great George Street house was home to John Trevanion, John Boyd's business partner and later his son-in-law.

The house in Great George Street brings us a little closer to Danson and its architect Robert Taylor, for, as Richard Garnier has noted, there are strong similarities between this development and a terrace of houses in John Street, Holborn, of the late 1750s, which is now accepted as Taylor's work.[7] Taylor was already popular with City merchants. His prominent new bank for Sir Robert Asgill, built in around 1756 at 70 Lombard Street (demolished in 1915), was the first of its kind in Britain, and shortly after he was appointed jointly with George Dance senior to repair and remodel London Bridge.[8]

John Boyd's fortune in the making was reflected in his move to Danson in 1753. He took on a full repairing lease on a well-

Figure 2.3
Lewisham House in Ladywell, Lewisham, in the late 1870s. Augustus Boyd took a lease on this building in 1743.
[By courtesy of Lewisham Local History Centre]

*Figure 2.4
Freeholds purchased by Boyd in 1759–84.
[© Crown Copyright and database right 2011. All rights reserved. Ordnance Survey Licence number 100019088]*

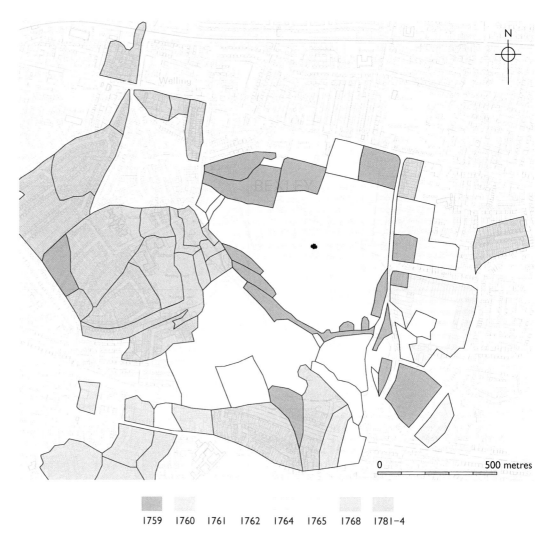

1759 1760 1761 1762 1764 1765 1768 1781–4

appointed house that was no more than eight years old with room for his growing family, extensive pleasure grounds and a dignified formal approach (*see* Fig 1.4). In 1759, Boyd paid £3,500 for the freehold of the property from Mary Styleman and the trustees of Styleman's bequest.[9] But Selwyn's Danson had an obvious fault. Located at the bottom of a hill on the eastern side of the estate, it offered no long views over the gently undulating Kentish landscape, and nor was it part of a view.

Towards the end of 1759 Boyd set about acquiring the freeholds of surrounding properties, whether large or small (Fig 2.4). In November, he paid £240 for 10 acres of land to the south, in Blendon, clearing it of cottages and farm buildings. There were two transactions for smaller properties to the south-west in 1760, each proving difficult to negotiate, but vital for consolidating the estate's boundaries. Two transactions in 1761 totalling £1000, secured him more than 65 acres, mostly of woodland, which formed the core of the park. The most important year for the development of the grounds was 1762 when he managed to obtain all the outstanding freeholds on the property, securing 180 acres to the north of the old house, taking in the ridge that bisects the estate and the house's future site. In the next two years there were small purchases, and then the pace of buying declined suddenly. Taken together, Boyd concluded 49 separate transactions between 1759 and 1765. Now that he had one of the largest freehold estates in Bexley, Boyd began to take part in parish affairs, and was elected to the parish vestry in December 1762.[10] A second period of growth followed between 1781 and 1784, with 26 transactions to achieve freeholds mostly outside the boundaries of the park, in areas now overlaid with interwar development. One of these was a purchase in 1781 of part of Great Chapel Field, an area south of the lake, which enabled Boyd to extend the park to its final form.

Boyd's decision to rebuild Danson would explain this period of freehold acquisition. In 1761 he notified the parish vestry that he intended to alter the terms of the charitable bequest, established by John Styleman's will of 1734, so that it could be satisfied by an annual payment of £100. The alteration required a private act of Parliament, which Boyd obtained in June 1762. One of the requirements of this was that he must spend at least £5,000 'in building a capital messuage and out-office' within five years.[11]

This is the only firm piece of documentary evidence to survive for the building of the house and the redesign of the landscape. There is no contemporary record of Boyd's engagement of an architect but he would probably have waited until he was sure of obtaining both the freehold of the larger part of the estate and the variation to the terms of Styleman's bequest before commissioning any designs. The most likely period for the initial design of the house and the landscape is therefore in the latter part of 1762.

A large, undated pen-and-ink drawing for Boyd's new parkland in the Bexley local studies collection shows the old canals and ponds consolidated into a serpentine lake, with an open expanse of turf to the south and north (Fig 2.5).[12]

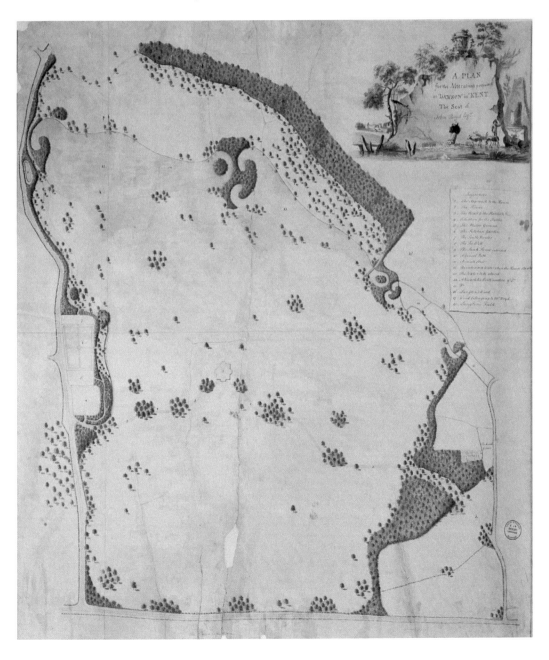

Figure 2.5
Nathaniel Richmond.
Proposed plan for the new park at Danson, c 1762–3.
North is at the bottom of the drawing. The larger elements of this scheme can be identified in the landscape today, especially those to the south of the house.
[By permission of Bexley Local Studies and Archive Centre, PEDAN/2/4. (DP094078)]

Dotted about the land are clumps of trees, and in the centre is a square-plan villa with projecting bays on the north, east and south sides. Here, in very broad terms, is the park that survives today. Despite the rather mannered, sinuous line that runs through the drawing, the underlying idea fits with the landscape tradition established by Lancelot 'Capability' Brown (1715–83). Indeed it is Brown who was credited with the design by several writers: Thomas Fisher in 1776, Hasted in 1797, and Edward Brayley in 1808.[13]

However, the Bexley drawing is in the hand of the landscape designer Nathaniel Richmond (1724–84).[14] Richmond had previously assisted Brown, very probably on the laying-out of Moor Park in Rickmansworth. He set up his own nursery in Lisson Green, Marylebone, in 1759, and by 1762 was designing the grounds at Stoke Park (Bucks). Richmond's landscape drawing, dating probably to 1762, has some bearing on the house, for it marks the villa's position at the centre of the new park exactly where it would come to be built, sitting astride the east–west ridge that bisects the landscape. The house is a villa – a square free-standing block without wings. Indeed, as will be shown below, the evidence of the fabric suggests that Danson House was begun as a free-standing block, although not exactly as shown in this plan.

According to the numbered legend on his drawing, instead of accommodating the stables in a service wing attached to the house, Richmond would have set them well away to the east with separate access from what is now Danson Road. Robert Morris had recommended a remote position for stables in his *Lectures on Architecture* of 1734,[15] and the important neo-Palladian houses of the 1720s – Wanstead, Stourhead, Marble Hill and Whitton Place – follow this pattern.[16] But in 1756, in *The Complete Body of Architecture*, Isaac Ware had already argued the contrary on the grounds of convenience: 'Beauty and use may be consulted together; and, instead of a plain square house, it will be possible, at a small advance in charge, to add wings to the centre and connect them by passages'.[17] It is doubtful that the architect of Danson House, Robert Taylor, had these words in mind when he came to finalise Boyd's design, but they do describe what happened after Richmond had prepared his drawing and the house sprouted service wings.[18]

In May 1763, the Revd Joseph Spence (1699–1788) visited Boyd to see how the park was progressing.[19] Spence is a curious figure in the history of the English garden. An Anglican priest now best remembered for his poetry, translations and literary anecdotes (he was friendly with Alexander Pope), Spence visited many estates later in life, trading on his wit and famous sociability while offering advice and comments on landscape design.[20] By the time he came to Danson he was much in demand, and his written notes and sketch plan of the estate and the house itself tell us a great deal (Figs 2.6 and 2.7 and *see* Appendix 1). First, Spence named Richmond as the designer in his comment on the view looking west:

W: View. Whether not too much blockt up by the Plant[ation] in Mr Richmond's Plan from Wellend Grove down to the little meadows.

Secondly, he recorded the position and a simple elevation of the new house in pen, and then added details of its north elevation in pencil (Figs 2.7 and 2.8). The central section of the north front shown in pencil had three arched openings at ground level instead of a flight of steps. The side bays are clearly shown

Figure 2.6
Reconstruction of the house in 1763, when Revd Joseph Spence drew his plan of the park (see Fig 2.7). The long canal, old house and Chapel House (see Figs 2.10 and 2.11) are shown in the background. By 1766 quadrant walls and service wings had been added (see Fig 5.1).

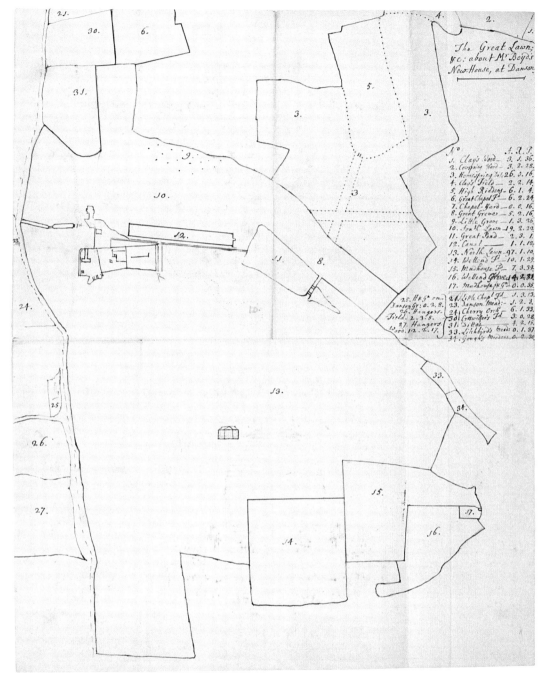

Figure 2.7
Revd Joseph Spence. Measured plan, in pen and brown ink over pencil, titled, 'The Great Lawn; &c: about Mr Boyd's New House, at Danson', and dated May 1763. North is at the bottom of the page (see Appendix 1). [Joseph Spence Papers. James Marshall and Marie-Louise Osborn Collection, Beinecke Rare Book and Manuscript Library, Yale University]

rising only as high as the bedroom floor. The arches at ground-floor level suggest that the perron (wide landing at the top of the stairs) had not been built, and perhaps not decided, at the time of Spence's visit. The perron as built differs from that shown in Malton's plan and view which were probably derived from drawings from Taylor's office. This implies two designs for the principal access to the house. If so, this aspect of the design had not been finalised in May 1763, although work on the

Figure 2.8
Detail of Spence's sketch of the north elevation of the house from his plan of May 1763 shown in Fig 2.7. [Joseph Spence Papers. James Marshall and Marie-Louise Osborn Collection, Beinecke Rare Book and Manuscript Library, Yale University]

Figure 2.9
Pen-and-wash sketch by the local antiquary Charles Thorpe of 'the Gothick Cottage, commonly call'd the Chapel-House', dated 1768. The cottage was built on land acquired by John Boyd in the village of Blendon, south of Danson, in November 1759. It lay beyond the area landscaped by Richmond and is the only one of the Park's earlier ancillary buildings to survive.
[© British Library Board MSS Add 32353 fol 247]

house itself had probably advanced to include the roof.

Spence's notes about the park on his sketch plan are less ambiguous, recording 'the Great Lawn; &c: about Mr. Boyd's New House at Danson' in the title at the top right-hand corner of the plan, and offering advice on the way the landscape should be laid out in a set of notes on two separate sheets. Here then, is a fairly firm date by which the first phase of construction must have been put in hand.[21] The 11 months between Boyd's Private Act of Parliament (June 1762) and this visit (May 1763) were long enough for work on the main block of the house to have been started. No house of this quality however, could be completed in all its details in such a short period, and certainly not for a mere £5,000. The family must have continued to live in the older house at the bottom of the hill, which Spence also shows on his plan north-east of '12', the 'Canal'. These same buildings, Styleman's house and grounds, are still present in a map of 1769 (see Fig 0.1).

According to a guidebook to London and its environs of 1782: 'The house presents itself to the view of every traveller, between ten and eleven milestones on the Dover Road'.[22] To the south of the lake, and beyond the lawn, just off Spence's plan, a cottage, imitating a church, with a spire and some Gothic windows, had been built by 1768 as a landscape folly to provide the right picturesque accent in the sweeping view to the south from the house (Fig 2.9).[23] The designer is not known but could well have been Taylor himself since he did on occasion work in the Gothic style.[24] The cottage survives at the junction between Blackfen Road and the Danson Underpass (Fig 2.10).

The ha-ha to the south of the house was essential to Spence's considered views. Although not shown on Richmond's or Spence's plans, it provided an inconspicuous barrier between the parkland and the gardens around the house. The ha-ha survived until the 1930s after which it was infilled and replaced by a beech hedge. Excavations in 1998 and 2003 revealed its construction: a nine-inch-thick wall, a mixture of Flemish and English bond, with a drainage course two courses above the

footings and occasional buttresses. It was restored in 2006 (Fig 2.11).²⁵

What is the status of Richmond's plan relative to Taylor's villa design? As completed, Danson was not a free-standing villa; it was a villa with wings, the building recorded in Barrett's portrait of 1766 and Malton's more architectural image published in 1792 (*see* Figs 0.2 and 0.4). Does Richmond's landscape design record an early stage in Taylor's thinking? Or was Richmond merely providing a generalised outline to satisfy his client's brief? And who decided where the new house should be sited in the first place, the architect, the landscape designer or even the client? After all, putting it in the middle of a large parcel of land, on the highest point with the best views, took no great design intelligence, and Palladio himself had advised that a villa should be in the centre of the estate, so 'that the owner can, without much difficulty, oversee and improve his lands around it'.²⁶

Figure 2.10 (left)
Chapel House, Blackfen Road, Bexleyheath, c 1760. Now separated from the park by the A2 dual carriageway, it was built as a roughcast cottage in the 'Gothick' style. The lead spire at the north end was an 'eye-catcher' that remained visible beyond Richmond's lake in views from the house. The chapel is marked on an estate plan of 1805 (see Fig 7.3).
[DP094091]

Figure 2.11
The house from the south-east in 2009. In the foreground is the restored 'ha-ha'.
[DP076293]

3

Robert Taylor, architect of Danson

Figure 3.1
Portrait of Robert Taylor, architect, attributed to William Miller.
[© National Portrait Gallery, London]

Robert Taylor (1714–88) was the architect chosen by John Boyd to build a new house at Danson. Taylor had begun his practice just as villas were being developed for smaller suburban seats in the 1750s.[1] He went on to become one of the capital's most successful architects and is now seen as a leading figure in the development of the suburban villa in the 18th century.[2] Danson is included in the suite of aquatint engravings commissioned by Michael Angelo Taylor from Thomas Malton to illustrate his father's work, published in 1792.[3] Malton's perspective view and plan provide the primary documentary evidence for the attribution of the design of the house with its wings to Robert Taylor (see Figs 0.3 and 0.4).[4]

Conventional wisdom has it that Taylor's success was due in part to his skill as a designer and in part to his great assiduousness. These attributes, combined with some canny investments, made Taylor a rich and influential man. He was knighted in 1782 on his election as Sheriff of London, and on his death six years later his estate was valued at £180,000, a vast sum for any professional to have amassed – far more than Sir Christopher Wren, James Gibbs or William Kent had achieved. He left the bulk of this fortune to Oxford University to establish 'a foundation for the teaching and improving of European languages'.[5] However, delayed by a dispute over the terms of the will, the Taylorian Institute was not actually built until the early 1840s to designs of C R Cockerell, the son of one of Taylor's own pupils. It is a measure of Taylor's high reputation that his obituary in *The Gentleman's Magazine* should have been written by Horace Walpole, who offered: 'he [Taylor] seems from the beginning to have been of those independent original powers which are reciprocally self-formed and self-forming'.[6]

A portrait of Taylor in the National Portrait Gallery from *c* 1782, when he was 64, shows a substantial and handsome man in the midst of life, with fine features and a penetrating gaze (Fig 3.1). He is in his achievement, knighted, and wearing the robes of the office of the Sheriff of London. Having transcended the accoutrements of his profession, he is no longer surrounded by plans, books, drawing implements, or references to ancient architecture; he is portrayed first and foremost as a gentleman of the City. Despite having enjoyed great success in his lifetime, later assessments have not been positive. It is hard to say why exactly, though it would appear merely an accident of history. Most of his securely attributed works in the capital have been demolished. No family papers survive and, worse, no office records. There is only a very small corpus of autograph drawings and, in addition to these, a checklist of some books that he owned, compiled posthumously. Taylor did not take any interest in 'public relations' and so never published his designs or thoughts on architecture in the way of more image-conscious architects such as Robert Adam or William Chambers. When Michael Angelo Taylor published the aquatint engravings of his father's work, only a few editions were printed. And so, perhaps, Sir John Summerson can be forgiven for not rating Taylor more highly in his now classic survey of British architecture, in which Taylor is passed over lightly, and dismissed as one of the 'conservative and consolidating' generation who began to practise in the 1750s. Taylor was, in his view, a Palladian architect with 'eccentric proclivities', favouring certain distinctive forms and using familiar motives in a 'personal way'.[7] Marcus Binney, who would take a keen interest in saving Danson, did more than any other modern writer to restore Taylor's reputation as a designer. He saw Taylor as a pioneer of neoclassicism, observing that his reputation had suffered because he was neither of Burlington's pioneering generation of Palladians nor part of Chambers' or Adam's neoclassical project.[8] In the end, this question should not cloud our appreciation of Taylor's buildings as architecture, and Danson is arguably one of his finest works.

Like many Georgian and early Victorian architects, Taylor was born into the building trades. His father, also Robert, was a master mason and monumental sculptor. The family lived at Woodford, Essex, in a villa built by the father. The son was born there and served an apprenticeship as a sculptor with Henry Cheere, after which the young Taylor made his way to Rome, it is thought, in the early 1740s. Very little is known about his time there, which was cut short by his father's death in 1742. Still, the fact that he took the trouble to go so far afield, at a time when Continental travel was still exceptional for artists, shows something of Taylor's high ambition and his desire to learn directly from the sources of good taste: classical and Renaissance art and architecture.

On his return, Taylor began the career for which he had trained. In England, in the 1740s, this effectively meant commemorative sculpture. His monument to Captain James Cornewall (1699–1747) stands in the west cloister of Westminster Abbey, an official state commission, coming directly from Parliament. By this time Taylor had been granted freedom of the Masons' Company, perhaps through his father's position in that body. Unquestionably, his most important early work was the commission in 1744 to carve the pediment of George Dance's Mansion House, opposite the Royal Exchange in the heart of the City of London (Fig 3.2). Though Taylor's father had been working on the building, he had to compete with leading lights Roubillac and his former master Henry Cheere, for the work. This pediment marked Taylor's arrival in London as a force to be reckoned with.

Binney identified Taylor's first architectural work at a house in St James's Square for Peter Du Cane, built from 1748 to 1750; following this was more work for the same client at Braxted Lodge in Essex from 1752 to 1756, described as 'repairs and rebuilding', and also some chimney pieces as well as furniture. Du Cane was a director of the Bank of England and the East India Company, and a number of Taylor's later patrons, including John Boyd at Danson, enjoyed similar positions. To quote

Figure 3.2
The Mansion House, City of London, (1739–52), by George Dance the Elder. Robert Taylor's first major commission on returning from Italy was carving the statuary in the pediment, beginning in 1744.
[© Richard Lea]

*Figure 3.3
Aquatint by Thomas Malton of the façade of Sir Thomas Asgill's bank, 70 Lombard Street, London, designed by Robert Taylor in about 1756. The bold, vermiculated rustication on the ground floor is characteristic of Taylor's work.
[By courtesy of the Trustees of Sir John Soane's Museum. (Malton 9)]*

*Figure 3.4 (opposite)
Comparative plans of villas designed by Robert Taylor. Danson and Purbrook House are shown without their service wings and quadrant walls; Sharpham without the attached older house, and Mount Clare without the portico added by Columbani, c 1780.*

Binney: 'Most of his [Taylor's] patrons were City men or had strong links with the City of London, and in some cases Taylor built them not only London houses or offices but suburban villas or country houses as well'.[9] Taylor was also unusual in taking on articled pupils, a practice that would become the norm in the early 19th century (though James Paine and John Carr are known to have done the same).[10]

After Braxted Lodge came three fine villas, Harleyford, Coptfold and Barlaston,[11] all being built in around 1755. There were also a number of important London commissions in quick succession: two houses in Lincoln's Inn Fields (Nos 35 and 36) finished in 1757, and the bank for Charles Asgill at No 70 Lombard Street (the first purpose-built one of its kind in this country) (Fig 3.3). There were several public works, the design of the Lord Mayor's Coach and the remodelling of Old London Bridge with George Dance the elder. These blossomed into larger-scale institutional appointments in the 1760s, notable among them the commission of Surveyor to the Bank of England, confirmed in 1764. In 1769 Taylor joined the government Office of Works and rose in successive posts until the office was reformed in 1782. He was surveyor to Lincoln's Inn, the Foundling Hospital, and Greenwich Hospital, though this instruction was received in the year of his death. There are public works outside London as well as a number of villas, chiefly Chute Lodge, Ludgershall (Wilts, c 1768), Purbrook House, Portsdown Hill (Hants, 1770, demolished 1829), Sharpham House (Devon, c 1770), and, most magnificent of all, Heveningham Hall (Suffolk, c 1780), finished by James Wyatt.

Documentation of Taylor's oeuvre has grown over the years. Thomas Malton's 32 aquatints show just 11 of Robert Taylor's commissions, hardly enough to account for his wealth, his glowing testimonials and the number of articled pupils in his office, which included both S P Cockerell and John Nash.[12] Howard Colvin listed 34 works including attributions in 1954.[13] In 1984, Marcus Binney had expanded the list to 51.[14] In 1995, Colvin revised his list to 72,[15] and in 2008 revised it up again to 85.[16] The gap in the historical record in recent years has made been made good by Richard Garnier. Taylor's output can now be seen to be considerable, especially when these lists count adjacent terraced properties, such as the 18 houses in John Street, as a single commission.[17] No doubt there will be more attributions to come but it is only now that we can begin to assess the scale of Taylor's practice and his influence.

Robert Taylor and the villa

Taylor's generation of architects was well equipped with published literature on classical Roman architecture and several of them had personal experience of travel in Italy. Like James Paine, John Carr, William Chambers and even Robert Adam at the beginning, Taylor continued to use Palladian detail but made his own advances, particularly in the development of massing and planning. The villas designed by this new wave of architects reflected changing patterns of use and new forms of socialising in refined society. In the late 17th century, great houses were organised as a series of interconnecting state rooms, arranged enfilade (with doors in line), all governed by a strict hierarchy and formality. The later 18th-century house was designed for more varied activities, and new pastimes: card playing, literary discussions, musical performances, dancing, and dining at different times. A circuit of rooms became the norm: a round of interconnecting

a) Harleyford Manor (1755)

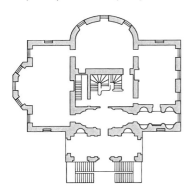

b) Coptfold Hall (soon after 1755)

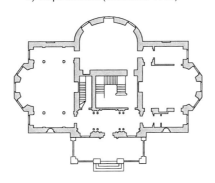

c) Barlaston Hall (1756–8)

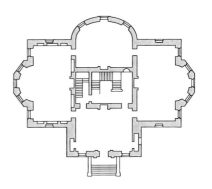

d) Asgill House (1761–4)

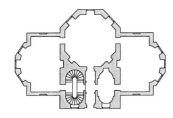

e) Danson (1762–6)

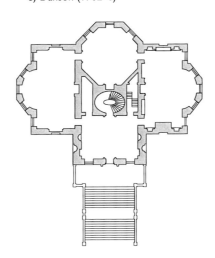

f) Chute lodge (c 1768)

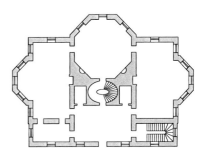

g) Purbrook House (1770)

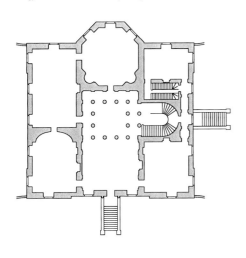

h) Sharpham House (c 1770)

i) Mount Clare (c 1771–2)

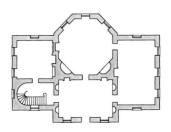

0   20 metres

*Figure 3.5
Harleyford Manor near Great Marlow, Buckinghamshire, from the north-east in 1910. It was built in the mid-1750s by Taylor, and restored as offices in the mid-1980s.
[© Country Life. L8082/8094]*

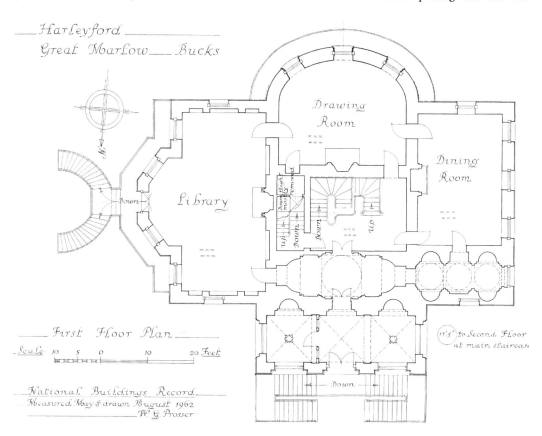

*Figure 3.6
Plan of the principal floor at Harleyford in 1962.
[MD63/00467]*

apartments that could be used in different ways.[18] This fashion emerged from Norfolk House in St James's Square, built in the 1750s by Matthew Brettingham (demolished in the 1930s).[19] There are many variations on this plan adapted to large and small houses.[20]

Danson offers such a concise expression of the arrangement that its plan has an almost diagrammatic quality, appearing more like a graphic device than an arrangement of masonry walls. Polygonal bays and octagonal rooms producing lively and crisply defined block plans characterise the work of this generation of architects. Richard Garnier has used the adjective 'crystalline' to describe the effect.[21]

Taylor's villas form a neat set of designs, following a definite evolutionary pattern, with development and continuity across time (Fig 3.4). Binney identified five main traits.[22] The first is that all Taylor's villas are astylar, that is, their exterior proportions are governed by the classical orders of architecture but do not have either giant columns or pilasters ('giant' referring to orders spanning two storeys or more). Also, they were designed to stand in splendid isolation in the landscape. Most have a compact plan, square or rectangular, with a pediment over the main entrance front. Their secondary and garden elevations are usually dominated by a curving or segmental bay window. Finally, all Taylor's villas, including Danson, have top-lit central stairs (at Asgill House and Mount Clare they are off-centre) and a circuit of rooms opening one into the other, quite often four principal rooms on the *piano nobile*, or principal first-floor level. We could also add that most of Taylor's villas have double-height kitchens and a vaulted passage, or dry area, running around the building, a feature which enabled the service areas to be fitted within the envelope of the house, on its lower levels.

The first three villas, Harleyford (1755), Coptfold (soon after 1755), and Barlaston (1756–8), were executed in brick, and form a neat group, in which one senses Taylor finding his way architecturally (*see* Fig 3.4a–c). Harleyford has a rather peculiar arrangement of rooms on the *piano nobile* (Figs 3.5 and 3.6). Instead of the great unity of form that characterises his villas of the 1760s, there is variety in the treatment of the spaces. The relationship of these rooms to the central stair (here square in plan) is not well resolved and the stair seems oversized for the house. By contrast, Barlaston's plan has the clarity we associate with Taylor's later works but the stair is still quite large and intrudes into the circuit (*see* Fig 3.4c). All these houses have large curved bays corresponding to an oval or D-plan room within. Their ornaments – whether the 'Chinese Chippendale' stair rails or decorative plaster reliefs – are shot through with a Rococo sensibility, light and gay, standing in contrast to Palladian austerity and clear geometry. The roof of each house rises behind a parapet.

Asgill (1761–4), Danson (1762–6), Chute (*c* 1768), and Sharpham (*c* 1770), all in stone, form another group, in which the curved bays are replaced by canted bays (*see* Fig 3.4d–f and h). Essentially, these stone-faced villas are austere in their underlying form, achieving simplicity of treatment and great immediacy of effect. This is particularly true of Asgill, whose principal floor is entered at ground level, anticipating a trend in late Georgian architecture (Fig 3.7). Binney notes the way that its roof springs directly from an eaves cornice, a treatment that suggests the detailing is related to the common rafters, harking back to the Roman architect Vitruvius' description of the features of an entablature being generated from timber constructional elements. Asgill is one of the smallest villas in the group, and smaller than those already discussed, which lends it an informal air that looks forward to more modest villas of the Regency and early Victorian periods. Taylor's skill as an architect shows itself in his manipulation of the relatively small internal volumes and the unpretentious main stair. The arrangement of rooms and axes within the building is complex without being laboured, so that overall a symmetry and sense of balance is achieved.

The same age as Boyd, Robert Taylor was, by 1762, a well-established architect with a highly developed speciality in the provision of villas for City business men. These qualities, coupled with Taylor's familiarity with Boyd's peers, made the choice of him as the architect for Danson almost inevitable.

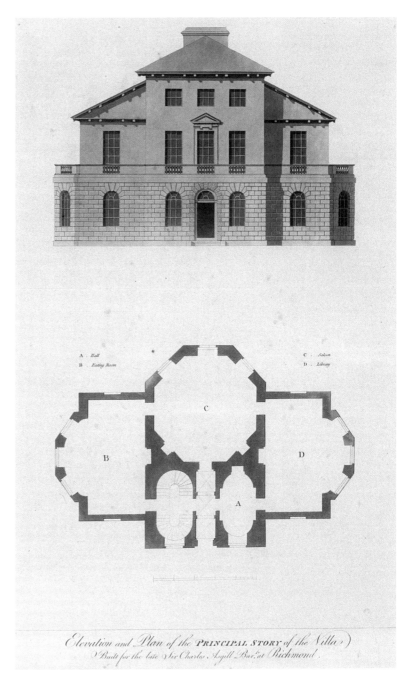

Figure 3.7
Thomas Malton. Aquatint engraving of Asgill House in Richmond, published in 1792, designed by Taylor, and built between 1761 and 1764.
[By courtesy of the Trustees of Sir John Soane's Museum. (Malton 10)]

4

Proportion and structure in Taylor's villa at Danson

Figure 4.1 (right)
The frontispiece of Robert Morris's Rural Architecture, *1750. The villa elevation illustrates the geometric basis of Morris's approach to design.*
[RIBA Library Photographs Collection]

Figure 4.2
Danson from the north-west in 2009, showing the canted bays.
[DP076263]

Danson was designed initially as a free-standing building with a central three-bay pedimented block, a form characteristic of the neo-Palladian villa in England from the 1720s onwards (*see* Figs 2.6 and 2.7). Robert Morris promoted this formula in 1750 in his *Rural Architecture Consisting of Regular Designs of Plans and Elevations for Buildings in the Country* (Fig 4.1). He supplied a proportional scheme, based on square and double-square proportions, which he indicated by circles drawn on the elevations.[1] Morris's plate no 2 shows a villa plan similar to that of Danson with a canted bay on the garden front, while plate no 30 demonstrates how such a bay might have been raised in a villa design with four nearly identical elevations.

Taylor adopted Morris's proportional schemes for the elevations of Danson House, but made greater use of the canted bay on the side and rear elevations (Fig 4.2). The canted bay broke up the mass of the rectilinear

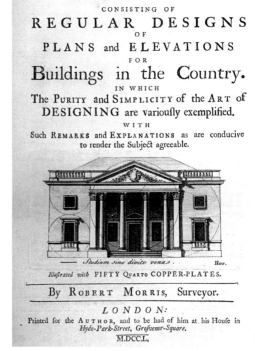

neo-Palladian villa and added interest to the interior by creating variety, space and a range of viewing points. From inside or out, the fragmentation of the rectilinear box aided the integration of the house into its surroundings, whether conceived of as an object in the landscape or as a viewing platform. Earlier, at Harleyford, Coptfold and Barlaston, he had favoured circular or elliptical bows, but from the 1760s he consistently employed the canted bay (*see* Fig 3.4a–c).[2] The canted bays on the side elevations were originally single storey, as at Asgill (*see* Fig 3.7), although with pitched roofs rather than balustraded platforms. This simple device – the roofed canted bay – which became ubiquitous in later Victorian terraced housing, seems to have been Taylor's innovation.

PROPORTION AND STRUCTURE IN TAYLOR'S VILLA AT DANSON

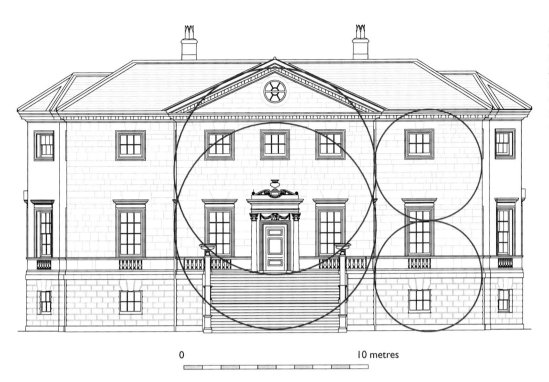

Figure 4.3
Proportional study of the main elevation of Danson House, illustrating Taylor's use of simple ratios – represented by circles – for the massing and openings of the façade.

Figure 4.4
The principal or north front of Danson House in 2009. In setting out the height of the middle sections of the house, Taylor used a proportional system that implied the existence of a giant column of the Ionic order.
[DP076958]

*Figure 4.5 (right)
Detail of Danson's eaves cornice.
[© Gordon Higgott]*

*Figure 4.6 (below)
Palladio's Ionic capital and entablature from Book I, p36 of* I Quattro Libri dell'Architettura *(1570). The cornice was the source for Taylor's eaves cornice at Danson.
[By courtesy of the Trustees of Sir John Soane's Museum]*

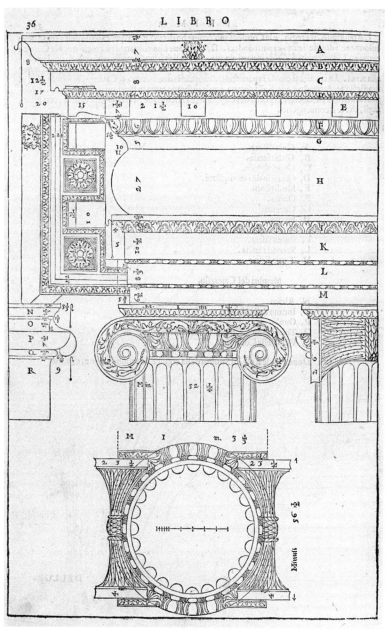

In approximate terms, Danson's front elevation conforms to the example on the frontispiece of Morris's book (Figs 4.3 and 4.4). The central section is governed by two circles, one from the ground floor to the lower edge of the cornice and the other from the base of the first floor to the sloping sides of the pediment. The recessed parts of the elevation are twice as high as they are wide, and the height of the elevation as a whole is twice its width. Square and double-square proportions are also used for the window openings (4 by 4 and 4 by 8 feet), as in Morris's illustrations.[3] Moreover, the gap between the window openings of the central projection is 6 feet, or one-and-a-half times the window module.

The elevations of Danson are treated in the classical manner but without a detached or engaged 'order' of columns or pilasters. Nevertheless, a ghostly rule overlays the design, and this dictates the size of the main eaves cornice in relation to the wall surface below. Against the smooth-faced wall one has to imagine missing giant columns or pilasters of the Ionic order, since the cornice at Danson is taken directly from Palladio's version of the Ionic cornice in Book I of his *Quattro Libri* (Figs 4.5 and 4.6). In Palladio's Ionic order, the column is 9 diameters (or modules) high, the entablature is one-fifth of the column's height, and the cornice five-twelfths the height of the entablature, a ratio which is also three-quarters of the column module.[4] Taylor's Ionic cornice at Danson is 2 feet high, implying a module of 2 feet 8 inches and a column height of 24 feet. If an Ionic column 24 feet high is applied to the wall above the rusticated ground floor, the entablature would be 4 feet 9½ inches high and its cornice would be 2 feet high. The total height of column and entablature would be 28 feet 9½ inches. This is only 3½ inches more than the built dimension of 28 feet 6 inches – sufficiently close to suggest that Taylor worked within Palladio's system when proportioning his external elevations.

The rooms on the principal floor follow proportions which had become standard in English domestic architecture by the mid-18th century and are based on classical precedents. The entrance hall is 26 feet 8 inches by 20 feet, a ratio of 4:3, while the library and dining room are both twice as long as they are wide (36 x 18 feet), a proportion that Vitruvius recommended for dining rooms.[5]

The masonry and timber structure

Danson's load-bearing structure is of brick with stone facing, the stone being an Oxfordshire limestone from the Wheatley or Headington quarries (Figs 4.7 and 4.8).[6] The internal partition walls on the basement, ground, and principal floors are of brick only; they change to timber on the second and attic floors, a system that is the product of considered design, and is very stable. Taylor's intentions for the wall elevations of the entrance hall, dining room and library were revealed during the restoration work. Each internal wall was formed in a series of round-arched recesses, flanked by rectangular recesses and, above these, roundels formed in brick, a composition that calls to mind the Roman triumphal arch (Fig 4.9). Comparison with his earlier villas at Barlaston and Harleyford suggests that Taylor must have intended a plaster lining that adhered closely to the underlying brickwork.[7] In the event, it was only in the entrance hall that he adhered strictly to his projected scheme; in the dining room and library this articulation was either ignored or altered when the house received its plaster lining.

English oak was used where the timber was either exposed to the weather or built into masonry. Thus, the lintels spanning all the door and window openings and the wall plates at the top of structural walls are all in oak, as is the timber set in the brickwork for the attachment of skirting, dados and cornices (Fig 4.10). The exterior frames of the sash windows are also in oak although the boxes and other internal parts are pine. The door frame for the front door is a composite of pine with an external facing of oak. Baltic pine was used for much of the internal structural framework.

Dendrochronological analysis has revealed that the latest timber to be used in the original construction of Danson was felled in 1758. It also revealed a close match with timber probably grown in what is now Belarus, and found in 1696 in Dannenstern House in Riga. This was the home of a merchant and ship builder, Ernst Metsue von Dannenstern, who is known to have exported timbers to Western Europe. Several timbers from this group bore the initials 'E+S' (Fig 4.11), probably denoting the timber yard in Riga that shipped the timber from its origins in the Dnieper Basin. The timber used to form the roof boarding, on the other hand, appears to have come from central-west Sweden or

Figure 4.7 (above)
A cutaway reconstruction of how the masonry carcase would have appeared in 1763–5, when the internal walls of the principal rooms were articulated with recessed arches, roundels and rectangular panels.

Figure 4.8 (left)
Looking north from within the canted west bay at bedroom-floor level during restoration, showing the stone-clad brick construction of the exterior wall, revealed when the canted bay was heightened in the early 1780s. The wrought-iron hanger in the foreground helps carry the bedroom floor below, and replaced the timber truss shown in Fig 4.15.
[B880014/51]

Figure 4.9
The north wall of the library, partially stripped of plaster, revealing the underlying masonry articulated with round arches, roundels and rectangular recesses.
[K961386]

Figure 4.10 (right)
A bedroom window during restoration in 1996. The bricks are laid to English bond and the surround is reinforced with limestone blocks and a timber lintel.
[F961557]

Figure 4.11 (far right)
A rafter end marked with the initials 'E+S' – the initials probably refer to a timber yard in Riga, Belarus.
[F960452]

Figure 4.12
Cutaway diagram illustrating the double-floor construction above the dining room, complete with trussed floor girder and soundproofing between the joists.

Norway. That used for the raising of the side bays at Danson suggests that the bays were raised 10 to 15 years afterwards, in the mid-1770s. This batch of timber came from another eastern Baltic source, probably a forest in northern Poland or further east. One of the rafters from this batch was branded with the initials 'FT'.[8]

Danson has 'double-floor' construction, typical of high-quality buildings of this period: one set of joists supports the floor, while the ceiling is suspended from another (Fig 4.12).

The design of each floor frame was varied according to its functional requirements (Fig 4.13). In the library, the long diagonal principal is a single timber, deriving support from the cross-walls below. At bedroom-floor level, the floor frames are in part suspended from king-post girders and trussed partitions (Figs 4.14, 4.15 and 4.16).

The complex plan of the attic-floor structure is determined by the roof, since the floor beams function as ties for the roof trusses (Figs 4.17 and 4.18). The trusses consist of principal rafters supported by queen posts that form the internal attic walls. The principal rafters rise to the flat section of the roof, which consists of yet another series of trussed girders. Around the dome, the lower parts of these girders were cut away when the ceiling was raised in the 19th century, surprisingly without disastrous consequences.

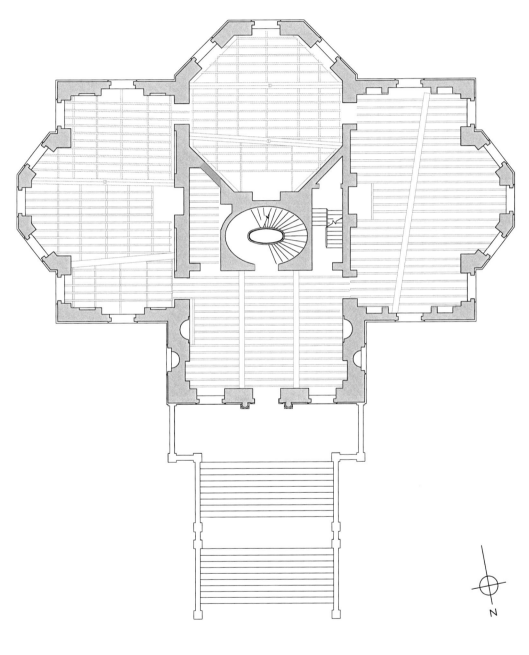

Fabric surviving from 1763–5
Conjectural reconstruction

0   10 metres

*Figure 4.13
Reconstruction plan of the masonry shell of the principal floor in 1763–5, before fitting out, showing the floor frames.*

Figure 4.14 (right) Plan of the bedroom floor as built, 1763–6, showing the masonry shell with the floor frame.

Figure 4.15 (below left) Reconstruction of the bedroom floor and partition walls as they might have appeared in 1763–6, showing the pairs of trussed girders that spanned the canted bays before they were raised to eaves level in the 1780s, by adding an extra storey (see Fig 7.4).

Figure 4.16 (below right) The paired trussed partitions that separate the two bedrooms over the library, photographed in 1996 after the addition of new timber studs. [F961530]

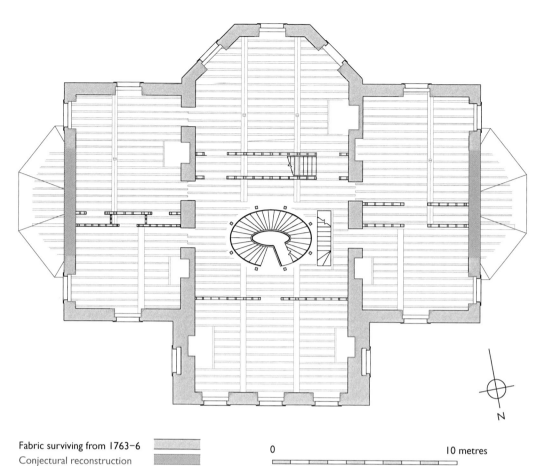

Fabric surviving from 1763–6
Conjectural reconstruction

0   10 metres

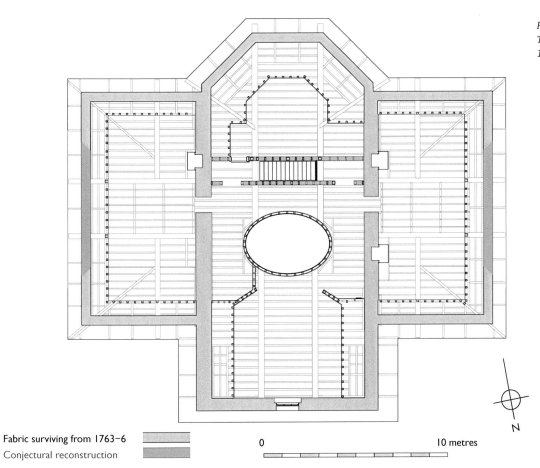

*Figure 4.17
The attic floor as built, 1763–6.*

Fabric surviving from 1763–6
Conjectural reconstruction

0   10 metres

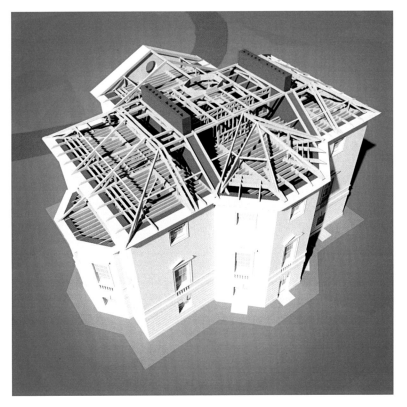

*Figure 4.18
Reconstruction of the roof structure, showing its relationship to the attic floor frame.*

5

Completing the house and landscaping the park, 1765–1773

When and why did Boyd's freestanding villa acquire its wings, and how long did it take to complete the landscaping of the park (Fig 5.1)? In George Barrett's painting of Danson House and the Boyd family from the north-west, the villa block is linked by curving quadrant walls to the two large, stone-built wings, square in plan, with prominent corner pavilions crowned by pyramidal roofs (*see* Fig 0.2). The wings can be seen in an engraving of 1794 (Fig 5.2), based on an earlier drawing by Richard Courbould, and Thomas Malton also includes them in both his plan and view of 1792 confirming Taylor's authorship (*see* Figs 0.3 and 0.4).[1]

Malton's plan identifies the west wing as a stable block. It has stalls for 12 horses in the north and south ranges and bays for three coaches against the east wall, facing into an open yard. Stairs in three of the towers suggest that they had upper floors, probably to accommodate the stable-hands. The south-east tower has a raised ground floor, though its purpose is unclear. The eastern wing, positioned on the kitchen side of the main house, was for 'Kitchen Offices'. It would have provided ancillary functions for baking, brewing and laundry. The open-sided circular structure in the south-east corner was probably an oven, indicating that this part of the kitchen wing was the bakehouse. The larger circular structure in the north wing could have been a water butt, or a brewing vat. Communication was at ground level, across the forecourt. One might have expected a concealed route, below ground, and indeed Sir John Soane mentioned two tunnels at Danson in his March 1815 Royal Academy Lecture (Fig 5.3),[2] but no trace of a tunnel has been found.

Nothing now survives, above ground, of these wings, and significantly, there are no scars to show where the quadrant forecourt walls were bonded to the walls of the main villa. The wings and their quadrant walls were thoroughly demolished when a large proportion of the masonry and many of the architectural features were reused in the present stable block built in 1805 (*see* Figs 7.9, 9.4 and 9.10 and *see* Chapter 7).[3]

Despite the paucity of documentation it is clear that the house was finished in 1766, within the time span set out in the private Act of Parliament obtained by Boyd in 1762. We know this from the fact that the dining room is lined with paintings by the French artist Charles Pavillon, one of which is signed and dated 1766. The organ in the library is also dated 1766 (*see* Chapter 6). Window tax returns show a jump in rateable value in 1766, suggesting that the house was finished in that year.[4]

We have seen how in Nathaniel Richmond's

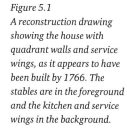

Figure 5.1
A reconstruction drawing showing the house with quadrant walls and service wings, as it appears to have been built by 1766. The stables are in the foreground and the kitchen and service wings in the background.

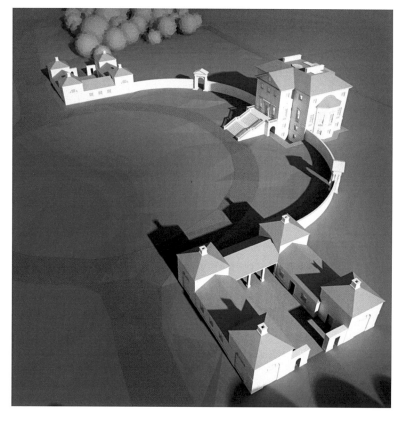

Figure 5.2
Engraving of Danson from the south-east, published in 1794, from an earlier drawing by Richard Corbould. The villa is shown with the side bays raised.
[A980203]

plan of 1762–3 and Joseph Spence's annotated drawing of May 1763, Boyd's villa is shown as a freestanding block in the centre of the park (*see* Figs 2.5 and 2.7). Richmond draws it square in plan rather than rectangular on the west–east axis, and does not show the deep entrance bay and front steps on the north side. Spence also draws the north elevation without the staircase, and indicates that its main entrance was at ground level, through a central arched door in the rusticated podium.

The evidence from the drawings and fabric points to Taylor having revised his designs for the building soon after the start of construction, adding the quadrant walls and wings, and perhaps also the staircase and perron. This supposition is borne out by two contemporary historical accounts, both of which suggest that the alterations were made while the work was progressing. Edward Hasted wrote in his authoritative *History of Kent* in 1797 how the house was altered and enlarged while it was being built, although he did not say whether Taylor himself was responsible for the changes:

> Boyd erected on an eminence, a quarter of a mile from the old seat, a most elegant mansion of Portland stone, the inside of which is decorated in a superb and magnificent taste, and gave it the name of Danson-hill. The original design for this structure was given by the late ingenious Mr. Taylor, architect of the Bank, but several alterations were found necessary to be made to it, for the accommodation of a family, whilst the house was building, and two wings were added to it for that purpose.[5]

Figure 5.3
An imaginary aerial perspective of Danson by the artist R D Chantrell, used to illustrate Sir John Soane's 'Lecture 9', delivered at the Royal Academy in March 1815. Chantrell omitted the quadrant walls, and Soane mistakenly assumed that the wings were connected to the main building by underground corridors, an arrangement he described as 'damp and inconvenient'.
[By courtesy of the Trustees of Sir John Soane's Museum. (18/5/1)]

Figure 5.4
Memorial to John Boyd's first wife, Mary Bumpstead (d 1763), in the south transept of Bath Abbey. It was probably designed by Taylor himself.
[© Richard Lea]

Hasted appears to have based his account on an earlier description by Thomas Fisher, written in 1776, about 10 years after the house was finished:

> At a small distance from Welling, on the south side of the road, is Danson-hill, upon which stands the seat of Sir John Boyd, baronet. The original design for this structure was given by the ingenious Mr. Taylor, well known from the great works executed by him at the Bank, but several alterations are said to have been made in the plan whilst this house was building.[6]

An explanation for the revisions to the design can be found in changes to Boyd's family circumstances soon after the private Act of Parliament in June 1762. In March 1763, John Boyd's wife Mary died in Bath at the age of 29. The elegant statuary memorial to her in the south transept of Bath Abbey testifies both to Boyd's developing taste in classical sculpture and to the involvement of Robert Taylor as sculptor, for the unusual motif of a console bracket carved with acanthus leaves is one he employed beneath the vaults in one of the four Transfer Offices at the Bank of England (1765–8) (Fig 5.4).[7] Mary's death may have held up the work, for Boyd's land transactions fell off at about this time. It is likely therefore that construction of the wings and front stairs did not begin until the latter part of 1763. The next significant date is August 1765, when Boyd's father Augustus died and John inherited the bulk of the estate. A year later, on 1 August 1766, he married Catherine Chapone, daughter of the Revd John Chapone of Charlton, Gloucestershire. A younger woman, she gave birth to

Figure 5.5 (right)
The main entrance, framed by attached Corinthian columns and a straight entablature. The 'pediment' of a vase flanked by scrolls was added by William Chambers, c 1768–70.
[DP094087]

Figure 5.6 (far right)
One of two side doors in the vestibule at Somerset House, Strand, London, designed by William Chambers, c 1776, and incorporating a scroll-block motif similar to that over the main entrance at Danson.
[© Richard Lea]

two more children. His inheritance and forthcoming remarriage may have provided Boyd with the means and motivation to revise the decorative treatment of the dining room, saloon and library, a process that was underway by June 1766, when the French painter Charles Pavillon was working in the house on his cycle of paintings for the dining room on the theme of Bacchus and the Golden Age (*see* chapter 6).

Although the exterior of Danson must have been complete in 1766, there is one telling difference of detail between the entrance front in Barrett's painting and the building today: Barrett does not show the scroll-block pediment on the entablature above the main entrance (Fig 5.5 and *see* Fig 0.2). Executed in creamy-white Portland stone, this scroll-block motif is clearly an addition to the door surround and the rest of the façade, all built from yellow-brown Oxfordshire stone. Stylistically the motif is attributable to the architect William Chambers, known from the fact that the two side doors of the vestibule at his Somerset House in the Strand, begun in 1776, have an almost identical feature over their cornices, although with a bust rather than a vase on the central plinth (Fig 5.6).[8] At Danson this motif reads as a signature. Moreover, the door surround itself, with Corinthian columns carrying a full entablature, and with swags of oak leaves draped over a central plaque, is based on a design for a Corinthian doorcase published by Chambers in his *Treatise on Civil Architecture* in 1759, where it is described as 'a composition of Inigo Jones' (Fig 5.7). Taylor amended this source by omitting the pediment, while Chambers, by adding a triangular-shaped overdoor sculpture, reinstated the original massing of the feature. He also sought to blend his addition with Taylor's doorcase by using an oak leaf swag in the centre detailed to match those in the frieze below. The vase at the top may be the one Chambers refers to in his design for a chimney piece in the dining room: 'to draw the Vase for Mr Boyd' (*see* p44 and Fig 6.22).

William Chambers and Danson

William Chambers' earliest involvement at Danson was probably in 1768 when Boyd paid for a painting of a landscape with a waterfall for the overmantel in the saloon by the French artist Claude-Joseph Vernet, whom Chambers had known in Paris in 1749–50. Chambers designed the frames of the two paintings over the doors in the saloon, one of which was a now lost view of the house from the south that was probably the work of the Swedish artist Elias Martin, who came to London from Vernet's studio in 1768.[9] Chambers also designed a new chimney piece for the dining room (*see* Figs 6.18 and 6.21). It is carefully integrated with the decorative scheme for this room and its design is probably contemporary with those for the picture frames. The chimney pieces in the saloon and library are also from designs by Chambers (*see* Figs 6.43 and 6.53).

In June 1770 Boyd commissioned Chambers to design a bridge and a temple in the grounds. Boyd wrote that the bridge was to be:

> ... of one arch 30 feet wide. The piers on Butments of brick and the bridge and rail of wood, not to rise too high, the Passage over 9 feet in the Clear and to bear a Carriage.'[10]

It was to span the newly formed lake at its narrow western end, where it probably replaced Thomas Wright's Chinese house and two bridges. The only visual record of the bridge is in Corbould's view of the house and lake, published in 1794 (*see* Fig 5.2). It shows a low, single-arched timber bridge, based on an example in Book III of Palladio's *Quattro Libri* (Fig 5.8), spanning the narrow west end of the lake. In his reply to Boyd's letter, Chambers referred to his designs for a bridge and a 'temple':

Figure 5.7
A design for a Corinthian doorcase from William Chambers', A Treatise on Civil Architecture, *1759 (facing p65, no 8), which has much in common with the entrance door at Danson.*

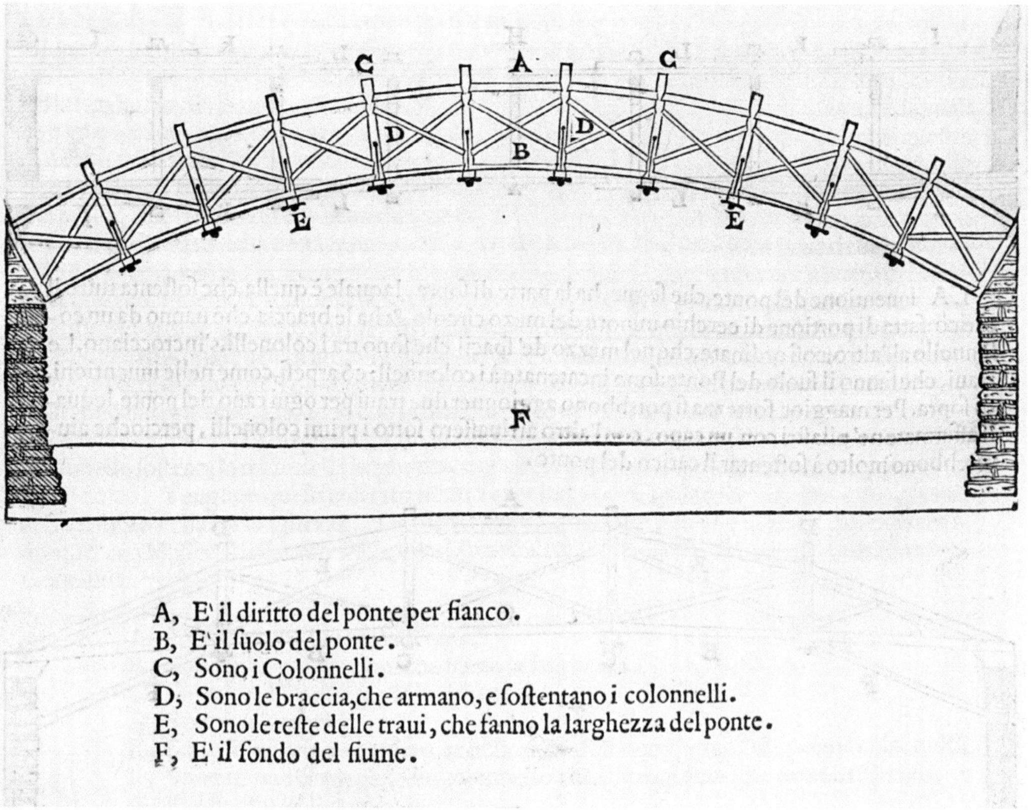

*Figure 5.8 (right)
Palladio's design for a timber bridge, from Book III, p18 of* I Quattro Libri dell'Architettura *(1570). It was the inspiration for the bridge at the west end of the lake designed by William Chambers, c 1770 (see Fig 5.2).
[By courtesy of the Trustees of Sir John Soane's Museum]*

A, E' il diritto del ponte per fianco.
B, E' il fuolo del ponte.
C, Sono i Colonnelli.
D, Sono le braccia, che armano, e foftentano i colonnelli.
E, Sono le tefte delle traui, che fanno la larghezza del ponte.
F, E' il fondo del fiume.

*Figure 5.9 (below)
The temple at the east end of the lake, designed by William Chambers, c 1770, and removed to St Paul's Walden Bury in Hertfordshire in 1961.
[By permission of Bexley Local Studies and Archive Centre. (B960795)]*

Excessive hurry ... has prevented my doing your designs. I am just a going to Hertfordshire for some days and on my return will immediately do your temple. In the mean time I have sent you a design of a Bridge wh[ich] has scarcely any rise at all. It is a thought of palladios and provided you have it framed by a skilful carpenter will do very well and look very handsome.[11]

This suggests that work on the new lake must then have been well in hand. It had not begun when Elias Martin painted his watercolour of the canal looking east towards the old house in 1768, but was probably finished by 1773, for in May that year Boyd wrote to Chambers to remind him 'that I am in your debt for the Drawings of my Temple, which is finished, and when your leisure will permit, should be glad you would come to see it, and spend a day at Danson with your Ladys'. Chambers replied: 'if the temple pleases you it will make me happy'.[12] He was referring to the Doric temple, which, until its removal to St Paul's Walden Bury in Hertfordshire in 1961, stood on the northern bank of the east end of the lake, and with steps leading down to the water's edge (Fig 5.9).[13]

6

Planning, decoration and iconography

Servicing the house: the basement floor

Taylor originally intended to house all the kitchen services in the villa itself. Only when the wings were added, towards the end of construction, were some of these functions accommodated outside the main house (Fig 6.1). In common with several of Taylor's houses, the kitchen itself is a double-height space, rising from a subterranean floor to half a storey above ground level (Fig 6.2).[1] Vaulted cooking spaces reduced fire risk and were made double-height to dissipate heat. The impressive high cross-vault over the kitchen is the product of a double-cube room with a canted bay in the centre of the long side. However, when first built, this room was not a single space. Brick

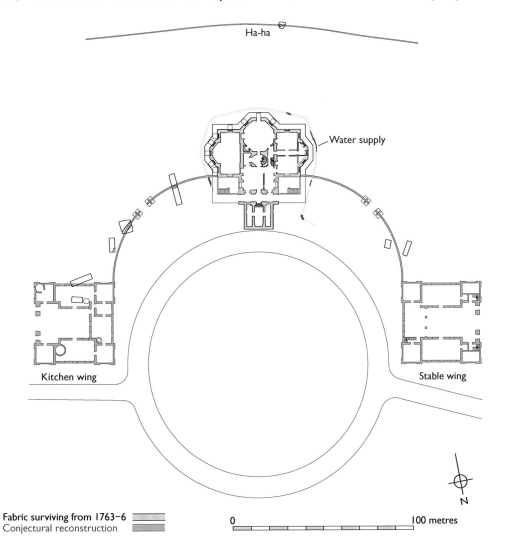

Figure 6.1
Plan of the ground floor of the house as built, 1763–6, with the wings reconstructed on the basis of Malton's plan (see Fig 0.3) and footings excavated in 1998. Water was ducted in a culvert around the south side of the house.

Figure 6.2
The vaulted double-height kitchen on the east side of the house.
[DP076966]

cross-walls spanned each end, rising half-height to the level of the ground floor. The space at the north end was close to the water supply in the open basement area at the front and probably housed a scullery, while the southern space was a bakery. Both rooms were probably roofed over to create mezzanine floors for dry storage (reached by internal staircases). The kitchen as a whole was well lit by windows in the end walls and in the projecting bay.[2]

The route from the kitchen to the ground floor was through the door in the wall on the west side, via a passage to the larder in the central octagonal room, and from here to the service staircase on the west side of the villa (Fig 6.3). The larder was for food storage and preparation, and was provided with large slate slabs set between the piers as cool work surfaces. To the west, the two windowless cellars were used for wine and beer storage. The central rectangular room to the north would have been the servants' hall. More vaulted cellarage was available under the perron and steps, where the northern three vaults were used for coal, served by coal holes in front of the main steps.

Before they were infilled in 1805, the basement areas either on side of the projecting frontispiece were open to the sky. Similar open basement areas survive intact at Harleyford and Barlaston (see Fig 3.4a and c). External access to the basement floor was probably provided in the basement area north of the kitchen, although no trace of a staircase is visible today.

The vaulted passage that surrounds the house at basement level is a common feature in high-quality building of the period (Figs 6.4 and 6.5). It separates the basement walls from the damp earth, and is a wise precaution on this site, where the subsoil is impermeable clay, with layers of gravel that act as aquifers.

A supply of spring water was delivered to the house through a brick conduit from somewhere near the surviving stable block.[3] It was ducted through a small brick culvert, about 9 inches in diameter (250mm), to a point north-west of the entrance front, from where it circled the house to the south, possibly supplying a tap in the cellar beneath the west bay. It then descended immediately east of the basement area on the north side of the kitchen, and passed beneath the present cellar floor (see Fig 6.1).

The ground floor

The ground floor is level with the external ground level and its rooms are slightly taller than those of the basement floor (see Figs 6.4 and 6.5). When planning this floor Taylor had to combine public spaces on the north and south sides with two private apartments on the west side, while orchestrating access to the floors above and below (of 'high' and 'low' status respectively) (Fig 6.6).

At Danson, rooms for private or family use on the ground floor and raised ground floor can be distinguished from service rooms by their

PLANNING, DECORATION AND ICONOGRAPHY

lath-and-plaster linings fixed to timber battens, nailed to the brickwork (*see* Figs 6.9 and 6.31). In the service areas on the basement and attic floors, by contrast, the plaster was applied directly to the brickwork (*see* Fig 6.2). The cavity between the linings and the brickwork made the polite rooms easy to heat and free from damp.

Interiors wholly lined from floor to ceiling with plaster were new in London houses around 1760. In the early 18th century, interiors tended to be fully panelled, while in the middle decades of the century, rooms often had plaster above a panelled timber dado. The transition was complete by 1774 when Thomas Skaife wrote: 'Wainscotting, in this refined age is quite obsolete, and seldom used, except in studies, or offices for servants, etc.'[4]

In Boyd's time, as today, the ground floor was entered through the brick-vaulted corridor

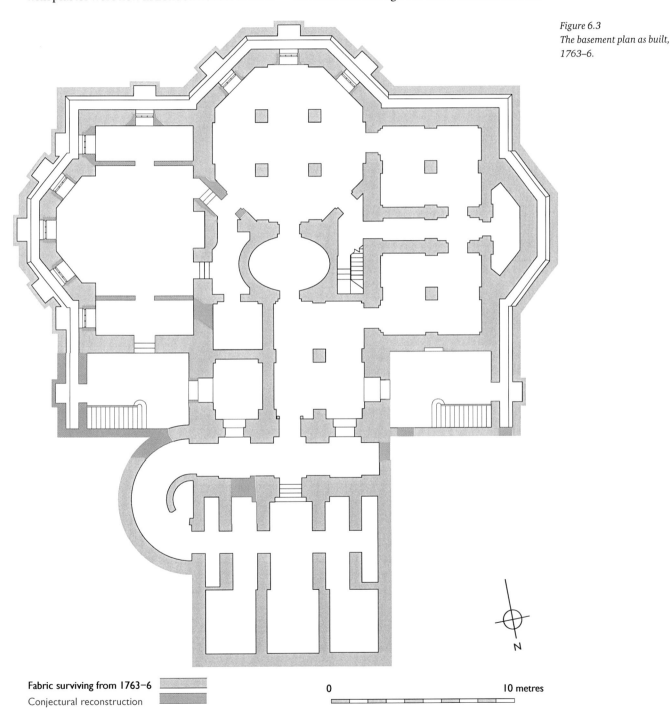

Figure 6.3
The basement plan as built, 1763–6.

Fabric surviving from 1763–6
Conjectural reconstruction

0   10 metres

Figure 6.4
Reconstructed east–west section through the house in 1763–5, before fitting out, showing the concealed masonry and timber structure, and the vaulted passage surrounding the house at basement level to protect the walls from damp earth.

Fabric surviving from 1763–5
Conjectural reconstruction

0   10 metres

Figure 6.5
Reconstructed east–west section showing the house as fitted out in 1766. The triumphal arch scheme has been abandoned in the dining room and library.

Fabric as fitted out in 1766
Conjectural reconstruction

0   10 metres

PLANNING, DECORATION AND ICONOGRAPHY

beneath the perron. This gave access to the house by three passages: the central passage – now a shop – led to the lowest stage of the main staircase and to the breakfast room beyond, the right passage led to the service stair and the left one to a closet on the left-hand side of the staircase.[5] The two rooms on the west side of the house, together with the closet in the window bay, probably served as bedrooms for family use. One was entered directly from the hall area and the other, to the south, from a door beside the servants' stair. Alternatively, one or both of these rooms could have been used as a servants' hall or housekeeper's room.[6] The corridors and rooms on the north side accommodated a variety of functions. This was the hub of the servicing of the house, its main day-to-day entrance, and the place where – under the watchful eyes of the butler and the housekeeper – the different strands of movement would have crossed most often.

The large chimney piece in the breakfast room may have been moved here from one of the principal rooms at first-floor level – perhaps

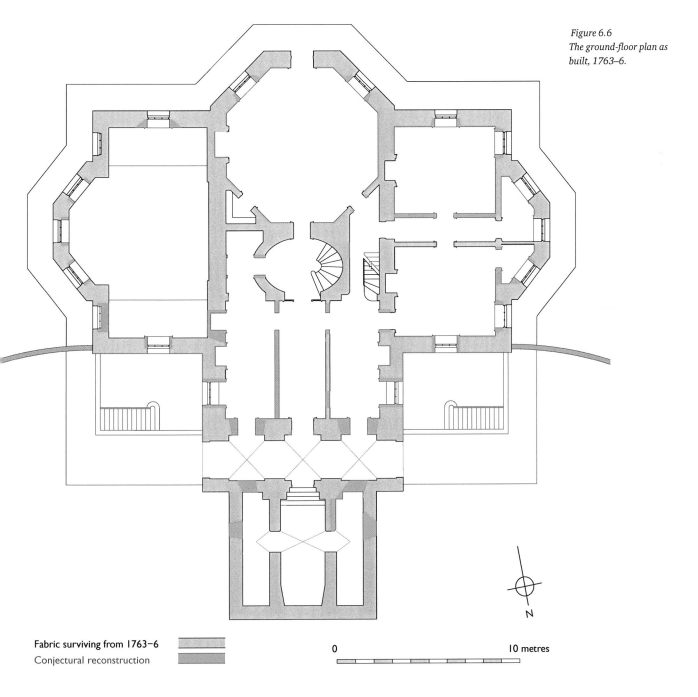

Figure 6.6
The ground-floor plan as built, 1763–6.

Figure 6.7
The chimney piece in the breakfast room on the south side of the ground floor.
[DP076963]

the library – when the three new chimney pieces designed by Chambers were installed in these rooms in about 1768 (Fig 6.7). It is in Oxfordshire limestone rather than marble, and is more architectural than figurative in character, with a large Greek meander ornament in the frieze and tapering fluted jambs surmounted by round paterae. Similar jambs are found on Taylor's chimney piece in the dining room at The Oaks, Carshalton, Surrey, *c* 1770, where there is also a sequence of plaster plaques identical to those in the Danson library (*see* Fig 6.52).[7]

Figure 6.8
Reconstructed plan of the principal floor or piano nobile *as built, 1763–6.*

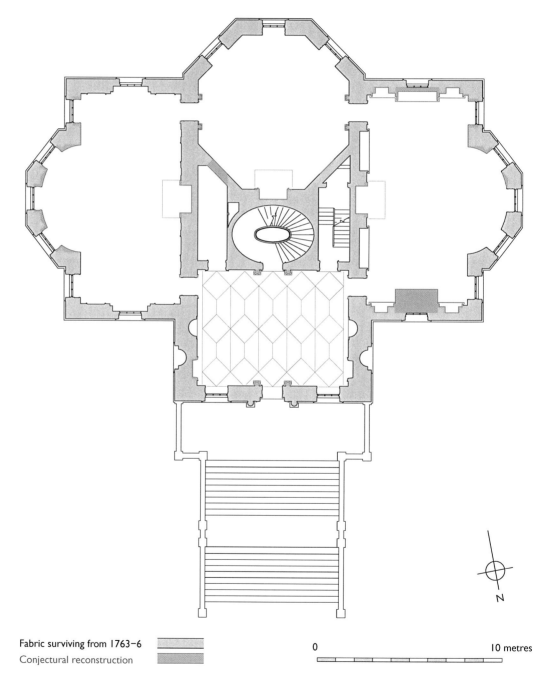

Fabric surviving from 1763–6
Conjectural reconstruction

0   10 metres

PLANNING, DECORATION AND ICONOGRAPHY

The *piano nobile* or raised ground floor

The *piano nobile* or raised ground floor of Danson is a simple circuit of four rooms, the spaces arranged tightly around an elliptical, top-lit stair, with the service stair occupying the rectangular compartment between the stair and the library on the west side and a closet in the corresponding space on the east (Fig 6.8 and *see* Fig 0.3).[8] As noted above, the dining room and the library embody changes in the treatment of their walls which indicate that Taylor revised his ideas for decorating these rooms at a late stage, probably in 1765–6. Restoration work in both rooms in 1998–9 revealed a 'triumphal arch' scheme of arch-headed, rectangular and circular recesses in the brickwork of end walls and similar roundels above the side windows and doors (*see* Fig 6.4). The Danson recesses would have housed busts, shelving or relief ornaments, as in the libraries at Harleyford (1755) and Barlaston Hall (1756–8) or the former ballroom at 4 Grafton Street (1770s).[9] In practice, however, it was only in the entrance hall that Taylor adhered to this severely classical scheme of interior decoration.

The entrance hall

The simplicity of treatment of the entrance hall answers this room's purpose as a threshold to the outside world. Paved in large, lozenge-shaped Portland stone slabs and lacking a fireplace, it is the most austere in the suite of rooms on the *piano nobile* (Figs 6.8, 6.9 and 6.10). Taylor has created perfect symmetry by repeating the doors to the dining room and library as cupboard openings on the opposite sides of the walls and using the same Ionic pedimented surround for both principal doors.

An entrance hall was traditionally a place for the display of sculpture, and Danson was no exception. The inventory of 1805 notes two Venus and Adonis groups in the entrance hall, both 'brought from Rome' (*see* Appendix 2), and the 1922 sale catalogue records portrait busts in the roundels.[10] The niches probably

Figure 6.9
The west wall of the entrance hall in 1995. Continuous vertical battens were nailed to the brickwork for fixing the plaster linings.
[Paul Dixon Photography]

Figure 6.10
The entrance hall in 2009.
[DP076971]

39

Figure 6.11
The dining room, looking north, in 2009.
[DP077010]

as much light as possible into this north-facing room is apparent from the unusually large glazed panel in the front door.[12] In the 1760s a pane of glass this size would have been expensive to produce and usually reserved only for mirrors. The final element of this interior is the sunburst ceiling rose, which appears to be original. Its hollow centre indicates that there would have been a hanging-light fixture, probably a brass lantern.[13]

The dining room and the cycle of paintings by Charles Pavillon

The dining room is identified on Malton's plan as the 'Eating Room' (*see* Fig 0.3). It is a light interior, illuminated by the morning sun and lined with a complete suite of allegorical wall paintings in gilded frames (Figs 6.11 and 6.12). The panelled plasterwork accommodating these paintings conceals the circular and rectangular recesses in the brick walls that represent Taylor's earlier intentions for the internal articulation of the end walls and the adjacent door and window bays. Only the arched recesses in the end walls were retained in the final scheme.

contained sculpture or vases, although no evidence of fixings survives. Paint research has revealed that the walls were in 'Naples yellow', a colour that matches the external stonework and was also used on the walls of the stairwell, thus creating continuity between exterior and interior throughout the principal circulation areas of the house.[11] Taylor's concern to admit

Figure 6.12
The south-west corner of the dining room, with the door leading to the saloon and the restored mirror in the arched recess.
[DP077016]

Figure 6.13 (above) Sarah Johnston's watercolour view of the dining room, looking north, c 1860. The original furnishings were removed from the property in 1806; those shown here belonged to the Johnston family. [© Bexley Heritage Trust]

Above the dado rail the plaster walls are formed into recessed panels with richly ornamented plaster frames of two sizes. Larger panels for figurative paintings have projecting frames with an entwined leaf pattern, fully gilded, while smaller panels for floral paintings have recessed moulding, partially (or 'party') gilded. The mirrors at each end of the dining room were installed in 2003–4 as replicas of originals shown in two of Sarah Johnston's watercolours of about 1860 (Figs 6.13 and 6.14). Their outlines had survived in the layers of paint applied to the walls, and fragments of the carved gilding were discovered in the Bexley local studies collection (Figs 6.15 and 6.16).

The important cycle of allegorical wall paintings was conceived and executed by the French artist Charles Pavillon in 1766, the date appearing beneath his signature on the painting of Apollo on the south wall (Fig 6.17 and *see* 6.28b). Pavillon was born at Aix-en-Provence in southern France in 1726 and is thought to have been the grandson of a goldsmith and engraver, Balthazar Pavillon. His work at Danson may be his earliest recorded activity in England. He was appointed master of the Trustees'

Figure 6.14 (right) Sarah Johnston's watercolour view of the recess at the south end of the dining room in about 1860, when both recesses retained their original timber gilded mirrors. [© Bexley Heritage Trust]

41

Figure 6.15 (above)
One of the recesses in the dining room during restoration in 1995. After the Victorian mirrors had been removed, the outlines of their Georgian antecedents were clearly visible in the wall paint. [© Richard Lea]

Figure 6.16 (above right)
Carved wood and gilded fragments of the original dining-room mirrors. [© Richard Lea]

Figure 6.17
Details from Pavillon's floral panels: (a) his signature and the date 1766, from the Apollo panel (see Fig 6.28b), and (b) the bullfinch at the foot of the hop vine, from the panel on the right of the chimney piece (see Fig 6.20c). [(a) DP077033; (b) DP077021]

Academy in Edinburgh in 1767, and exhibited there in 1770 with the Society of Artists. Despite living in Scotland, he kept up his contacts in London and was recorded as living at 'Mr. Inge's in Covent Garden' in 1768 and 1770, when he exhibited at the Royal Academy. He died in Edinburgh in 1772.[14]

The Danson paintings were removed from the house in 1971 and have since been cleaned and restored by English Heritage. The 19 paintings divide into 2 types: large panels depicting nearly life-size mythological characters, alternating with narrow panels filled with pictures of flowers. It is an ambitious and well-executed scheme, especially interesting for the unusual iconographic programme which relates not just to the function of the room but, more significantly, to the client's background.

Each picture was executed in oil on canvas, which was then glued directly to the wall's skim coat of lime plaster with animal glue.[15] The artist probably painted them on easels in the house while staying as a guest, for in June 1766 Boyd mentioned in a letter that the painter was then working in the house.[16] Pavillon would not have started until the dirty work of finishing the interiors was done. In any case, it is right to think of these pictures as 'finishing touches', and they were completed just in time, for Boyd was to remarry in August of the same year.

In their overall design, the floral panels have affinities with those used in tapestries from the Gobelins workshops in Aix, the artist's birthplace. It may be, then, that the artist spent some time in the French royal tapestry factory. There are also parallels with the work of the Italian painter Francesco Zuccarelli (1702–88), a contemporary of Pavillon, who had links with the Gobelins factory and spent time in England in 1752–62 and again in 1765–71 to capitalise on the increasing demand for decorative paintings in houses.

No precise classical source for the cycle has yet been identified, but the first two books of Virgil's *Georgics* appear to have informed the programme.[17] These are hymns of praise to Ceres and Bacchus, who appear in several guises in the Danson paintings. Ceres, the goddess of the harvest, was credited with bringing mankind to a Golden Age, and Bacchus, the god of the vine, with bringing civilisation to the ancient Greeks. The primary association of the two, and hence their suitability for a dining room, is with plenty and pleasure and, also, with love, because without Ceres and Bacchus *amor frigit*, or 'love grows cold', as Ovid wrote in *The Arts of Love*.

The focus of the cycle is the square panel over the chimney piece (Fig 6.18 and *see* frontispiece). This depicts a sacrifice to Bacchus, who is represented by a marble bust mounted on an altar. His presence is also felt in the overdoor picture to the right, where a pair of cupids holds a tambourine and pan pipes (Fig 6.19). These are standard emblems in depictions of the Triumph of Bacchus. The apples in this painting refer to the full-size figure of Pomona to the right of the fireplace (*see* Fig 6.26b), who is associated with the harvest and hence a consort of Ceres (*see* Fig 6.27b). In the opposite overdoor are two more cupids (*see* Fig 6.19).

One pours water onto tender shoots sprouting from a large terracotta pot, indicating spring, and echoing a theme announced at the main entrance of the house, where a delicate low-relief cameo-type carving depicts the head of a young woman girded with a wreath of flowers. There are more references to Bacchus either side of the chimney piece in floral panels that depict a fruiting grapevine on the left, and hops, or the English grape, on the right (Fig 6.20b

Figure 6.18
The dining room chimney piece by Chambers c 1768 and the overmantel painting, A Sacrifice to Bacchus, by Charles Pavillon, 1766, with the decorative panels shown in Fig 6.20b and c to the left and right.
[DP076255]

Figure 6.19
The dining-room overdoor paintings, by Pavillon, 1766: (a - left door) cupids with floral tributes; (b - right door) cupids with tambourine, pan pipes and a basket of grapes.
[© Bexley Heritage Trust]

and c). Robert Adam associated the two plants in precisely this way in the dining-room ceiling at Osterley Park which he probably designed in 1762.[18]

The chimney piece was designed by William Chambers and inserted soon after the decorative scheme was finished, probably in about 1768 (Fig 6.21). Chambers' original drawing for the chimney piece – inscribed 'Eating Room' – is a preliminary design, altered in execution, and is the only one of the three designs for chimney pieces connected with Danson, purposely designed by the architect's office for a room in the house (Fig 6.22). Chambers wrote on his design, 'to draw the Vase for Mr Boyd', but neither his drawing nor the vase he proposed for the chimney piece have come down to us. It is possible that the vase in question is the one in the centre of the composition above the main entrance door (*see* Fig 5.5).

The side pilasters of the chimney piece are adorned with a Bacchic 'thyrsus', a composite symbol of a rod of fennel capped with a pine cone wound with ribbons, hanging from a leopard's mouth. The block above each leopard's head is carved with an Apollo's lyre, and the frieze is ornamented with swags of grapevines, again a reference to Bacchus. They have been varied from those in the design, and the central roundel was changed to a tablet with a relief adapted from a well-known classical source: a frieze panel of the Choragic Monument of Lysicrates in Athens, published by James Stuart in his *Antiquities of Athens* in 1762 (Figs 6.23 and 6.24).[19] The Danson version shows Bacchus reclining on the hide of a goat, in place of the drapery on the source engraving (goats being another element of the Bacchic triumph), holding a wine bowl and stroking the head of a leopard. A stump of a grapevine reinforces the Bacchic theme. Chambers used the same form of tablet in a chimney piece carved by Joseph Wilton for the drawing room of Peper Harow House in about 1765.[20] The quality of the carving of the Danson example is extremely fine and could also be by Wilton (Fig 6.25) as he was one of the most accomplished decorative sculptors of his day and often worked for Chambers.

The story of Pomona, to the right of the chimney piece, and her pendant opposite, Vertumnus, is told in Ovid's *Metamorphoses* (Fig 6.26).[21] Vertumnus, a demigod smitten by the young Pomona, disguised himself as an old woman in order to praise his own virtues, but it was only when he bravely revealed his own identity that Pomona reciprocated his love. The Danson pictures show the moment of truth, the unmasking, a scene of revelation taking place across the Bacchic altar with a fireplace below. Depicted here, in allegorical terms, is a wedding engagement, culminating in a wedding feast.

The story of Vertumnus and Pomona was a common subject in Italian, Dutch and French painting from the 16th century onwards. Pictures generally show Vertumnus in his disguise as a wrinkled crone leering over Pomona, who is often depicted as a naked, voluptuous young woman. In Dutch art, this is not meant as an incitement to lasciviousness,

Figure 6.20
Floral panels on the chimney piece wall in the dining room, by Pavillon, 1766: (a - to right of left door) exotic fruits, including a gourd, quinces and artichokes; (b - left of chimney piece) a grapevine; (c - right of chimney piece) a hop vine; (d - to left of right door) exotic fruits, including pomegranates, lemons and peaches.
[© Bexley Heritage Trust]

Figure 6.21 (top left)
The chimney piece in the dining room, designed by William Chambers, c 1768.
[DP076256]

Figure 6.22 (left)
Design by Chambers for the chimney piece for the dining room, inscribed 'Eating room' and 'to draw the Vase for Mr Boyd', c 1768. Pen and brown ink, brush and washes. Sheet 11¹⁵⁄₁₆ x 8in. The Metropolitan Museum of Art, New York, The Elisha Whittlesey Collection, 1949. [© 2010. The Metropolitan Museum of Art/Art Resource/SCALA, Florence]

Figure 6.23 (above top)
Detail of central plaque from the chimney piece in the dining room depicting the god Bacchus stroking the head of a leopard (see *Fig 6.24*).
[DP077014]

Figure 6.24 (above)
James Stuart and Nicholas Revett. Engraving of the Bacchus relief from the Choragic Monument of Lysicrates, Athens, in The Antiquities of Athens (1762).
[RIBA Library Photographs Collection]

Figure 6.25 (above left)
Detail of a leopard's head on the chimney piece in the dining room. The quality of the carving suggests the hand of the sculptor Joseph Wilton, with whom Chambers frequently collaborated.
[DP094336]

but rather a condemnation of lust. There is nothing odd about Pomona in the context of a dining room, given her link with the harvest and therefore with food and plenty. Vertumnus' unmasking is harder to fathom until one considers the circumstances of the commission. For, like Vertumnus, Boyd was an older man, 48 in 1766, courting a younger woman, his wife-to-be, Catherine. The fable is a deeply romantic one: truthfulness counts for more than physical beauty.

On the north wall of the dining room is Ceres herself, with heavy eyes, her whole frame sensuously loose and languid, exhausted it seems

Figure 6.26
(a) Vertumnus unmasking himself to Pomona and (b) Pomona, by Pavillon, 1766. Panels on the left and right sides of the chimney piece.
[© Bexley Heritage Trust]

Figure 6.27 (above)
(a) Bacchus *and (b)* Ceres, *by Pavillon, 1766. Panels on the north wall of the dining room.*
[© Bexley Heritage Trust]

Figure 6.28 (above right)
(a) Euterpia *and (b)* Apollo, *by Pavillon, 1766.*
[© Bexley Heritage Trust]

by the heat of the summer harvest signalled by an upturned rake (Fig 6.27). Bacchus is on the same wall, on the other side of the round-arched recess. He appears as a lithe young man, smiling and holding a ewer. Next to him is a krater, an ancient Greek wine vessel, and at his feet is a basket and ribbon, the latter a reference to his wand, the thyrsus, a standard element of the triumph that is echoed by the chimney piece. On the south wall, echoing Bacchus and Ceres, are, on the right, Apollo, god of the sun and patron of the arts, and on the left a less familiar figure, Euterpia, the Muse of lyric poetry, a genre with which Apollo is sometimes associated (Fig 6.28). The globe at Apollo's feet represents the harmony of the spheres, and the water is almost certainly a reference to the Hippocrene spring, which flowed from the spot on Mount Parnassus where Bellerophon, his winged horse, alighted. There, on Parnassus, Apollo won a poetry competition held among the gods. It may seem to be a series of unconnected subjects, but the themes – love, marriage, fecundity and the arts – are complementary, implying a kind of cultural efflorescence that in retrospect is entirely appropriate to this period of English architecture and applied art. The 1760s was a decade of great achievement, understood as such at the time. Here, then, is a room that celebrates achievement in the widest sense, radiating splendour, plenty and fruition.

The painted fruits and flowers are obviously appropriate to a dining room because of their sensual associations (Fig 6.29 and *see* Fig 6.20).[22] It would be surprising, in a scheme so carefully considered, if the fruits and flowers did not also relate to the adjacent painted figures, and indeed this turns out to be the case. Celia Fisher has identified three groupings in these side panels: first, plants with a long iconographic tradition in the arts (grapes, apples and pears); second, those which are unusual in paintings (hops); and, third, a small number of exotics (lemons, pineapples and pomegranates).[23] Some of the fruits relate to the function of the room, others to the figures in the paintings. The rose (in three panels) is traditionally an attribute of Venus, the goddess of love, and associated with poetry. Quince (three in one panel, to the right of the left-hand door), is also associated in Greek myth with Venus, so too is the pomegranate (in two panels), also known as the 'love apple'. Here it occurs with the pineapple, whose form echoes the pinecone, and hence refers back to Bacchus and his thyrsus, the pinecone-topped rod that he used to whip his female followers into frenzy.

Most of these species were known in England from the 17th century, but a few would have been recent introductions. Hydrangeas were American, brought in during the 1750s, as well as wisteria, chrysanthemum and dahlias. The latter were only successfully nurtured at Kew in the 1790s, which suggests that the Danson artist had access to botanical illustrations. This accounts for the accuracy of their depiction.

The 1805 inventory contains very little about the furnishing and other decorations in the dining room, noting only 'an oriental spar cistern on a marble splinth [sic]' carried by '3 bronze figures' and a 'fine French urn' for this (*see* Appendix 2).[24] There were two small pieces of statuary sculpture supported on matching high plinths and these appear in one of Sarah Johnston's watercolours from about 1860 (*see* Fig 6.13). The room is certainly large enough to have held a sizeable dining-room table but, in this period, it was unusual to have a permanent full-size table. Chairs and a table were commonly brought in for particular events. Indeed, the early 19th-century inventory refers to this room as the 'Parlour', which suggests its use was not restricted to eating but that it possibly also served as a morning room, because of its eastern aspect. This use persisted into the 19th century.[25]

Figure 6.29
Floral panels on the window wall in the dining room by Pavillon, 1766. The broader panels, (a) and (f), flank the canted bay, and the narrower ones, (b)–(e), are within the bay. The flowers include roses, honeysuckle and chyrsanthemums.
[© Bexley Heritage Trust]

The saloon as a setting for landscape painting

While the dining room was the most ambitious interior, with the greatest allegorical content, the saloon was designed to glitter and exude luxury (Fig 6.30). The large cornice, based on Palladio's Corinthian, is the most elaborate in the house, and is set above a plaster anthemion frieze decorated in mid-blue oil paint with the ornaments picked out in party gilding, a technique that was also used for the arabesque in the eight radial panels of the ceiling (Figs 6.31 and 6.32).[26] The walls are lined with plaster on lath, set flush with a set of timber battens that enclose each stretch of wall, showing that the room was designed for a rich wallpaper or fabric. Common practice in the hanging of wallpaper or fabric was to line the walls with scrim, calico, or hessian, stretched across the wall plaster, and tacked to the battens. Surviving rows of nails and strands of scrim attached to the battens show that the room was indeed finished in this way (*see* Fig 6.31). Paper or silk fabric was then attached to the scrim with glue.

The sale catalogue of 1805 refers to 'Walls hung with Blue Silk Damask'.[27] Sarah Johnston's watercolour of the saloon in about 1860 appears to show this (*see* Fig 6.32). Wallpaper at the time often imitated damask and the physical evidence is not sufficient to determine which was used in this room. The wall face above the fireplace preserves the pattern of the original wall-covering in the form of a residue on the plaster, presumably glue (Fig 6.33). Although the residue provides only a partial image of the pattern recorded by Sarah Johnston, it does show that the design included swags and urns, patterned in a half-drop

Figure 6.30
The saloon in 2009. The frames of the mirrors are replicas based on what can be seen in Sarah Johnston's watercolour (see Fig 6.32). The painting above the fireplace is a photographic copy of the original in the Walters Art Museum, Baltimore, USA (see Fig 6.35).
[DP076984]

PLANNING, DECORATION AND ICONOGRAPHY

repeat, and that the rolls of fabric were 26-inches wide.

The present wallpaper, manufactured by Adelphi and hung in 2003,[28] is hand-block printed to a pattern from a fragment of 18th-century paper in the Cooper Hewitt Museum where the design draws on a set of Chinoiserie motifs drawn and published by Jean Baptiste Pillement in 1773.[29] The colour matches the original blue ground of the frieze. There is no fabric or documentary evidence to suggest that there was a Chinoiserie element in the Danson wall-covering, but it is known that the octagonal drawing room at Taylor's Ottershaw Park, near Chertsey, built soon after 1761, was decorated in the Chinese style.[30]

The plaster lining of this room is largely as Taylor conceived it, except for an adjustment he made above the two doorcases. During restoration work in 1998, triangular gaps were discovered in the plaster above the entablatures, suggesting that Taylor first intended pediments in these positions. There is no sign that they were ever installed, and their omission can be explained by the decision to hang paintings above the doors while the rooms were being fitted out in late 1766 and early 1767 (Fig 6.34).

We first hear of the overdoor paintings in February 1767, when Boyd was commissioning the French artist Claude-Joseph Vernet (1714–89) to paint the 'Landscape with waterfall' for the wall above the chimney piece. The commission

Residue from damask
Damask roll widths
Conjectural reconstruction

Figure 6.31 (above left)
The saloon cornice in 1995. The exposed timber framework was the mount for a backing of scrim or hessian, for a wallcovering of paper or fabric.
[F931078]

Figure 6.32 (above right)
Watercolour view by Sarah Johnston of the west side of the saloon, looking into the library, c 1860. The mirror and picture frame (shown partially) were probably original. The blue wallcovering may be the one mentioned in an early 19th-century inventory.
[© Bexley Heritage Trust]

Figure 6.33 (left)
A drawn record of Taylor's original fireplace opening in the saloon, and the impression made by the original wallpaper on the wall above.

Figure 6.34 (above) The west wall of the saloon, looking towards the library, with George Barrett's painting above the door. [DP077004]

Figure 6.35 (opposite) Landscape with waterfall and figures. Claude-Joseph Vernet, French, 1768, oil on canvas, 69 3/8 x 53 1/4 in. John Boyd commissioned the painting in December 1766; it arrived at Danson in March or April 1768. [© The Walters Art Museum, Baltimore, USA]

is documented in two separate notes in Vernet's account book which record Boyd's original order for the painting in December 1766 and a revision on 24 February 1767, requiring a change in the dimensions of the painting:

[Order number 229] 'A painting for Mr Boyd, Englishman in London, ordered by Mr Vanloo, via a letter that he received from Mr Pavillon. It must be five feet wide by six feet and one inch high in English feet. It must include a large waterfall, some distant views and plenty of figures; the price is one hundred and fifty louis or 3600 l. I promised it for the month of March, 1768. It was ordered in December 1766'.

[Receipt number 137] 'By letter from Mr Pavillon of 24th of February 1767, written to Mr Vanloo, he has sent new measurements for the painting above, which are 5 feet 10 inches in height by 4 feet 6 inches in English feet, or 5 feet 5 inches 10 lines in height, by 4 feet 2 inches 3 lines in French feet, it must

still have some waterfalls with a seascape in the background. The two overdoors in the saloon where this painting is to hang, being landscape. All of the rest, with regard to the said painting, as above'.[31]

Vernet's painting – a superb example of the 'classical romantic' landscape in the manner of Claude Lorraine – was delivered to Danson in April 1768 [32] and remained in the saloon until it was sold at auction in 1805 (Fig 6.35). It now hangs in the Walters Art Museum in Baltimore, Maryland, USA, and a photographic reproduction takes its place in the restored saloon.

It is possible that the courier who delivered Vernet's painting to Danson was the young Swedish painter Elias Martin (1739–1818). He was studying in Vernet's studio at the Ecole des Beaux Arts in Paris in 1766 and was in London two years later exhibiting paintings at the Society of Artists, one of which was a watercolour entitled 'A View of Danson in Kent'.[33] The watercolour was almost certainly a study for a painting listed in the sale of Boyd's collection at Coxe, Burrell and Foster on 7 and 8 May 1805, as *A Picturesque View of the Back elevation of Danson, in the County of Kent, with Cattle and Figures*.[34] This study must be contemporary with Martin's sketch of the old house, seen from the west end of the canal (*see* Fig 1.2).[35] It is clear from the title alone that the *Picturesque View of the Back elevation of Danson* was meant to complement Barrett's view, described in the same sale catalogue as *A Perspective View of the Principal Front of Danson, in the County of Kent, with Lawn, Castle and Figures*. The prices confirm that they were viewed as a pair: £12 12s for the Barrett, £12 1s for the *Back elevation*. These two landscapes must have hung above the doors in the saloon, for the Barrett painting fits one of the spaces, and the other was considered its pair.[36] Thus, on each side of the saloon were views of Danson House, one from the front, another from the rear, while on the central rear wall, above the chimney piece, was Vernet's idealised landscape, opposite a real but nonetheless man-made landscape seen through the windows of the bay. Standing in the middle of this room, John Boyd would have been at the focal point of the landscaped park, and his pleasure in his creation was reflected in the paintings he chose.

The dimensions of Vernet's painting correspond closely to those in Boyd's revised order of February 1767,[37] and his request to reduce the size of the painting from 5 feet to 4 feet

Figure 6.36
William Chambers.
Preliminary design for the frames of the overdoor paintings in the saloon, with a clam-shell motif at the corner. [By courtesy of the Trustees of Sir John Soane's Museum. (42/3/8r)]

Figure 6.37
William Chambers. Alternative design for the frame in Fig 6.36, on the back of the same sheet. The design is for a simpler, flatter frame, with a goat's skull and acanthus motif at the corner. [By courtesy of the Trustees of Sir John Soane's Museum. (42/3/8v)]

Figure 6.38
John Yenn (draughtsman). Sketch for a modified version of Chambers' design in Fig 6.36. [©V & A Images/Victoria and Albert Museum, London (7078.2)]

Figure 6.39
John Yenn (draughtsman). Presentation drawing for a picture frame, based on the preparatory sketch in Fig 6.38. [©V & A Images/Victoria and Albert Museum, London (E.4984-1910)]

Figure 6.40
John Yenn (draughtsman). A more developed version of the design in Fig 6.39, with variations in the ornaments and gilding. This design was probably used to frame Vernet's painting over the chimney piece. [©V & A Images/Victoria and Albert Museum, London (3861.19)]

Figure 6.41
John Yenn (draughtsman). Presentation drawing of alternative design for the overdoor frame in Fig 6.37. Sarah Johnston's watercolour shows that this design was used for the overdoor frames (see Fig 6.32). [©V & A Images/Victoria and Albert Museum, London (3861.20)]

6 inches may have been made to ensure that the picture, with its frame, aligned with the existing fire surround.[38]

Evidence that Boyd engaged Chambers to work on the interiors of his house around the time that Vernet's painting was delivered in 1768 is found in designs for picture frames by Chambers and his draughtsman, John Yenn. Chambers' drawings are preparatory studies for alternative versions of the overdoor frames on two sides of a single sheet in the Sir John Soane's Museum (Figs 6.36 and 6.37).[39] The more elaborate design on the front of the sheet is inscribed, 'Profile & ornaments for the frames over Mr Boyd's doors', with an instruction above the drawing (presumably for Yenn) to 'lengthen out the ornaments …', while the design on the back is simpler, without a moulded profile and with a goat's head and acanthus motif in the corner rather than a clam shell. Yenn's drawings are on four further sheets at the Victoria and Albert Museum (Figs 6.38, 6.39, 6.40 and 6.41).[40] He followed Chambers' instruction to lengthen out the ornaments of the clam-shell design in a pencil study (see Fig 6.38) and rendered this as a neat pen-and-wash drawing, inscribed 'Design for a Picture Frame' (see Fig 6.39). He then produced a more carefully finished version of this design, in which he narrowed the frame from 5 inches to 4 inches, pulled in the ornaments from the inner corner, and added all the party gilding (see Fig 6.40). This was presumably the executed version of the design.

In Sarah Johnston's watercolour of the western door of the saloon, looking into the library, the bottom of the frame is visible (see Fig 6.32). It is decorated with the acanthus and goat's skull motif and lacks any indication of the deeply moulded profile of the more elaborate 'clam shell' version of the design. It would seem, then, that Boyd chose the plainer design for the overdoor frames. The more ornate 'clam shell' version was probably used to frame Vernet's painting over the chimney piece and has been adopted for the restoration of this feature.

The saloon chimney piece is Chambers' tour de force which, in design and installation, post-dates the completion of the room (Fig 6.42). A design for a chimney piece very similar to this one survives in a group of Chambers' drawings for interiors at Marlborough House in London in 1771–4 in the Avery Architectural Library, New York (Fig 6.43).[41] Boyd probably

PLANNING, DECORATION AND ICONOGRAPHY

Figure 6.42 (far left)
The saloon chimney piece.
[DP077002]

Figure 6.43 (left)
William Chambers. Design for a chimney piece at Marlborough House, London, c 1771–4. Chambers reused this design for the chimney piece in the saloon.
[Avery Architectural and Fine Arts Library, Columbia University]

Figure 6.44 (far left)
Central plaque on the saloon chimney piece, with a relief depicting the marriage of Cupid and Psyche.
[DP077003]

Figure 6.45 (left)
Bernard de Montfaucon. Engraving of the 'Marlborough Gem', depicting the Marriage of Cupid and Psyche, published in 1719.
[Royal Academy of Arts, London]

procured the design from Chambers in the early 1770s, when the architect was engaged in work on the grounds. The tablet at the centre of its frieze bears a relief of the Marriage of Cupid and Psyche, based probably on an engraving of a relief known as the 'Marlborough Gem', published and described by Bernard de Montfaucon in 1719 (Figs 6.44 and 6.45).[42] The scene is from the popular Renaissance allegory.[43] Cupid has his head covered to signify his hidden identity as a god, while Psyche, a mortal, is completely shrouded, denoting her ignorance. Chained together, they approach the marriage couch, and an attendant carries a sacramental torch as homage to Hymen, while another has a tray of fruit, symbolising fertility.

The library

The library depends for its effect on a stark contrast between the white finish of the plaster and joinery and the combination of a dark verdigris wall colour and the rich, dark mahogany used for the bookcases and organ (Fig 6.46).[44] As in the dining room, Taylor covered the recessed roundels and panels he had built into the brick carcase of the room to create a new decorative scheme (Fig 6.47 and *see* Figs 4.9).

Decorative schemes incorporating a suite of bookcases were not unusual in the mid-18th century. Chambers, for instance, designed a comparable bookcase for the Earl of Pembroke's villa in Whitehall. Made by Thomas Chippendale in 1760, it survives in Wilton House.[45] Sir Laurence Dundas commissioned a pair from Robert Adam for his house in the West End, 19 Arlington Street, again made by Chippendale.[46] Chippendale himself did a great deal to popularise this type of bookcase, publishing designs in his widely consulted *Gentlemen's and Cabinet-Maker's Director* of 1753.[47] This has 13 designs for bookcases in all, 3 of which are close to the pedimented bookcase on the south wall of the Danson library. Unlike Danson, however, all his designs call for glazed doors.[48] Batty Langley also published a design for a Tuscan-style bookcase similar to that fixed to the south wall of the library at Danson in 1756.[49]

On the east wall are three low-relief plaster plaques, the smaller elliptical ones above the doors serving as pendants to the larger central one above the chimney piece (*see* Fig 6.46).[50] In the outer plaques are standing maidens, the north one holding fruit and representing Pomona, and the south one holding a bound sheath of corn and a scythe, and denoting Ceres

55

Figure 6.46 (above) The library looking north. In the restoration of this interior, the roundel above the fireplace was reinstated, the organ returned, and the Victorian glass-fronted doors removed from the bookcases (see Fig 6.56). [DP076973]

Figure 6.47 (right) The library, looking south. As on the north wall, the recessed arches, panels and roundels in the brick carcase were covered over in the fitting-out. [© Gordon Higgott]

PLANNING, DECORATION AND ICONOGRAPHY

Figure 6.48 (top left) The plaque above the door from the library to the entrance hall, representing Pomona. [DP076981]

Figure 6.49 (top middle) The plaque above the door from the library to the saloon, representing Ceres. [DP076977]

Figure 6.50 (far left) The library chimney piece and overmantel plaque. The chimney piece was designed by William Chambers, and the plaque has a relief scene from the Roman Marriage Ceremony, *taken from an engraving by Pietro Santi Bartoli (see Fig 6.51). [DP076979]*

Figure 6.51 (top right) Engraving by Pietro Santi Bartoli from Amiranda Romanarum Antiquitatum *(1693), which illustrate decoratives found on ancient urns, vases, sarcophagi, wall-paintings, tablets and monuments in Rome. [Royal Academy of Arts, London]*

Figure 6.52 (middle right) The dining room at the Oaks, Carshalton, Surrey, added by Taylor in c 1770, but demolished in 1957–60. It had identical relief plaques to those in the library at Danson, although set within arches. Note also the similarity between the chimney piece and the one in the breakfast room at Danson (see Fig 6.7). [AA58/106]

Figure 6.53 (bottom right) William Chambers. Design for a chimney piece for General Ralph Burton, Haworth Hall, Hull, c 1763–8, similar to the one in the library at Danson. [Avery Architectural and Fine Arts Library, Columbia University]

(Figs 6.48 and 6.49). They face the woman in the central plaque, who is sitting on a couch having her feet washed (Fig 6.50). This is a scene from the *Roman Marriage Ceremony*, the third in a suite of four images engraved by Pietro Santi Bartoli and published as part of *Romanarum Admiranda Monumenta* (Rome, 1693) (Fig 6.51).[51] All three plaques reprise the themes of love and marriage, harvest and plenty already seen in the dining room. As Taylor used an almost identical group of round wall plaques in the dining room at The Oaks at Carshalton c 1770 (Fig 6.52), and there is no evidence for the plaques having been added to the walls after the 1760s, there can be little doubt that these relief ornaments are part of Taylor's original decorative scheme.

The library chimney piece is from a design which Chambers had used previously for a room at General Ralph Burton's house at Haworth Hall in Hull, although Boyd would have known of it from a version that Chambers published in his *Treatise on Civil Architecture* in 1759 (Fig 6.53).[52]

The organ provides further evidence for the completion of the interiors by Taylor in 1766, for its console has an ivory label inscribed 'Old England fecit 1766', and the outer bays of the case are carried on acanthus leaf consoles which can be associated with Taylor's work at the Bank of England (Figs 6.54, 6.55 and 6.56).[53] This inscription identifies the instrument as the work of George England (d 1773), one of the most skilled and highly regarded organ builders of the day. Domestic organs of this period that can be played and survive in their original context, are rare, especially in houses open to the public.[54]

Figure 6.54 (right)
The organ in the library, dated 1766 on the console.
[DP076974]

Figure 6.55 (far right)
An acanthus-leaf console supporting the organ pipes. It resembles the console on the wall monument to Mary Boyd (see Fig 5.4), and suggests Taylor's hand as designer of the organ case.
[© Gordon Higgott]

Figure 6.56 (below)
Watercolour view by Sarah Johnston of the north-west corner of the library and part of the organ, c 1860. The bookcases do not have the glass-fronted doors added in the late Victorian period.
[© Bexley Heritage Trust]

Before the 1750s, chamber organs were generally not found in houses. Most musical entertainment in the home took the form of suites and sonatas played on the spinet or harpsichord.[55] Handel's fame accounts for the change in taste. His organ concerti were hugely popular at pleasure grounds in the 1730s and 1740s. Later, organ concerts were given at Carlisle House in Soho Square, the Pantheon in Oxford Street and in Hanover Square, and other organ composers capitalised on Handel's success. As a result, after 1760, the date of publication of Handel's organ music, small organs began to be installed in domestic interiors, sometimes in specially designed 'Music Rooms' but also, commonly, in rooms that had another function, such as halls and libraries.[56]

The stairwell and the bedroom floor

The oval grand stair runs up to a landing on the bedroom floor where an Ionic colonnade rests on the plinths of the balustrade and supports a dome with a central skylight (Figs 6.57 and 6.58). The dome is timber and divided into pinewood panels painted in trompe l'oeil coffering in which rosettes alternate with thunderbolt motifs. Opposite the top of the staircase a central door once opened into a large, rectangular room lit by three square windows beneath the pediment (the single windows in the flank elevations were always dummies (Fig 6.59 and *see* Fig 0.5)). That the wall was originally designed to accommodate a central door is clear from the positions of the diagonal braces (Fig 6.60). This north-facing room, set directly above the entrance hall, may have been one of the public rooms, the family sitting room, or Boyd's study, and would have been hung with paintings. To the east of the stairwell is a pair of bedrooms, the southern of which gave access to the principal bedroom, and perhaps served as a boudoir or dressing room for the lady of the house. To the west is another pair of rooms. They once shared an interconnecting door and a cupboard space, and could have been separate bedrooms or a bedroom and dressing room. The walls of the landing were painted in Naples yellow like the entrance hall, while the timber columns, decorative mouldings and wrought-iron balustrade were painted in off-white oil.

Most of the light enters the stairwell through the dome's top row of glazed panels. Although analogies with the Pantheon and other domes would suggest that the top row of panels would originally have been panelled and painted like those below, this was not the case. Careful examination of the dome's structure revealed that the ribs around the top row of panels were of oak, cut to receive flat panes of glass, not pine and not curved panels, as in the lower parts of the structure. Furthermore, the paint on the glazing ribs was shown to be one with the trompe l'oeil scheme. Although none of the original glass survived, it was probably frosted or acid-etched like the circular windows lower down the stairwell that light the service stair and closet.

Taylor's wrought-iron staircase balustrade between the principal floor and the bedroom floor was heightened in the 19th century by the addition of the wrought-iron circular motifs immediately below the handrail. Its original

junction with the base of the column at landing level can still be seen as an outline in the paint. The present balustrade at the top of the service stair is a restoration. It was reconstructed from the evidence of mortices in the floor around the service stair, mortices in the column bases around the main stair and the outlines of the handrail preserved in paint.[57]

The bedroom floor presents a unified sequence of rooms in which differences in functions and status were articulated in their plaster cornices. The principal bedroom on the south side has the same enriched cornice as the stairwell, while in the room to the north, one of these mouldings – the dentil course – was left blank. In the ante-room or boudoir to the principal bedroom on the south-east side only the

Figure 6.57 (above)
The elliptical cantilevered staircase, looking up from the landing of the principal floor to the dome above the Ionic colonnade.
[DP076965]

Figure 6.58 (left)
The landing at bedroom-floor level. A circuit of eight Ionic columns supports an elliptical timber dome. The panels are painted in trompe l'oeil as stone coffers, ornamented alternately with thunderbolts and flower motifs.
[DP077028]

dentil course was elaborated. The remaining rooms on this floor have the same simple cavetto and cyma cornices found in the rooms on the ground floor. Throughout the bedroom floor, as on the ground floor, the original window-shutter boxes were square to the window. The shutters were removed and replaced with splayed reveals in the 19th century. The chimney pieces are mostly original although some, like the elegant example in the eastern half of the northern room, have been brought in from another room in the house and altered (Fig 6.61).

The attic or garret floor

The ascent to the garret floor was by the secondary stair between the landing and the large south bedroom (*see* Fig 4.17). The garret floor was probably shared between servants and family members. The floor plan is built around the walls containing the chimney flues, the oval domed roof light, and the secondary staircase. The room to the front of the house was lit by the oculus in the pediment; the other rooms were lit either directly or indirectly by roof lights set in the flat sections of the roof.

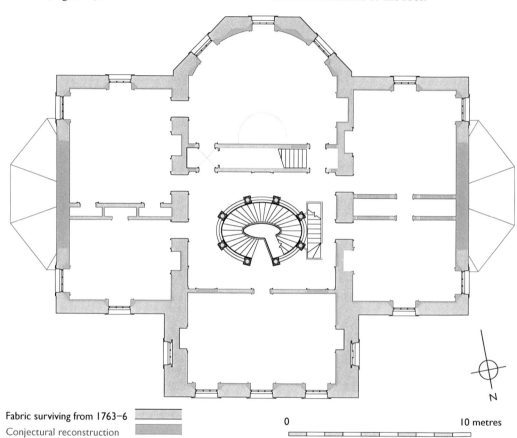

*Figure 6.59
Reconstructed plan of the bedroom floor as built, 1763–6.*

*Figure 6.60 (right)
The trussed partition to the north of the stair in 1995. It was built with a central door facing the top of the stairs. This was later blocked and new doors were formed at either end by cutting through the diagonal struts. [F952743]*

*Figure 6.61 (far right)
The fireplace, now in the eastern of the two rooms (originally one) above the entrance hall.
[© Richard Lea]*

7

The cost of life as a gentleman

Expanding and managing the estate

Boyd's business was built on international trade, and his library at Danson was filled with prints of harbours and ports, county maps, geological specimens from around the world and scientific instruments. It is perhaps not surprising to discover that in 1775, when he was created a baronet, he allowed himself a tour of the Continent.[1] He returned in 1776 with an ancient stone vase, known as the Piranesi Vase (Fig 7.1). Boyd purchased this extraordinary assemblage of ancient fragments from the painter and excavator Gavin Hamilton (1723–98) during his continental tour. The fragments of which the vase is composed had been excavated in 1772–3 at Monte Cagnolo, and then restored, but by whom is not certain.[2] Since the 1860s it has been in the collection of the British Museum.

The piece was greatly admired by those who saw it, and Giovanni Battista Piranesi published three large engravings of it in *Vasi …*, one with the following note (Fig 7.2):

> A perspective view of an ancient marble vase and its pedestal that can now be seen in England in the villa of John Boyd. It was discovered in excavations on the site of Hadrian's villa in 1769.[3]

Although there are precedents for vases in the antique manner in Palladian houses from this period (Adam's ante-room at Syon House for one), it is by no means certain that Boyd installed it in the house (the entrance hall being the only space that could reasonably have accommodated it). The marble is very large, 9 feet (2.7m) high, more than half the height of the hall. It weighs about 4 tonnes and there is no evidence for floor strengthening in the hall, compression marks or damage in the flooring.[4] Also, 'nella Villa del Sig. Giovanni Boyd' could just as easily mean 'at the villa' as 'in the villa', and it is ill-advised to treat Piranesi's inscription uncritically. By 1805, according to the hand-written inventory (*see* Appendix 2), this 'capital antique Vase supported by 3 figures' was standing in the 'Green House'. The location of this greenhouse is not known, since it does not appear on the estate plan of 1805 (Fig 7.3). Wherever the Piranesi vase was kept before that date, its present good condition argues that it had not suffered from prolonged exposure to the elements.

Soon after Boyd's return to England in May 1776, the American War of Independence began and the sugar trade dried up. Early in 1779 he was devastated by the death of his favourite daughter Mary Jane. He began to grieve again for the younger son, Augustus,

Figure 7.1 (below right) The vase purchased by Boyd during his Italian journey of 1775, the year he celebrated his baronetcy. Known as the 'Piranesi Vase', it was a contemporary assemblage of antique fragments. In 1868 John Johnston sold it to the British Museum where it is now displayed. [© Trustees of the British Museum]

Figure 7.2 (below left) One of three plates illustrating the 'Piranesi Vase' in Giovanni Battista Piranesi's Vasi … *(1778). [By courtesy of the Trustees of Sir John Soane's Museum]*

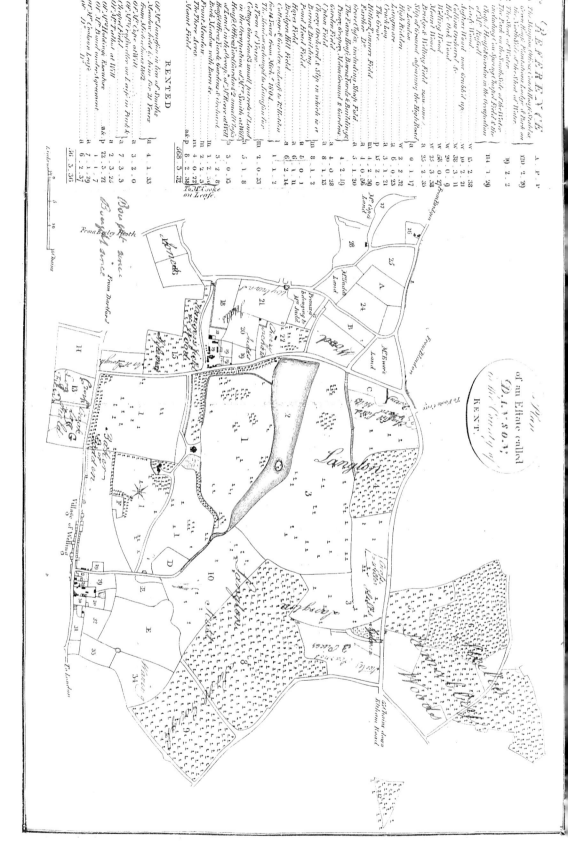

Figure 7.3
Plan of the estate published with the sale particulars in 1805. North is at the bottom of the sheet. On the east (left) side of the estate is Danson Farm (numbered '19'). At the top of the plan, beneath the oval legend, and opposite the road 'To Foots Cray', is Chapel House, set in an apse-ended enclosure (see Figs 2.9 and 2.10).
[By permission of Bexley Local Studies and Archive Centre, CODAN/1]

who had died from fever in 1773.⁵ His mood was black and inconsolable. His son John sympathised, but only to a point. Boyd junior had his grandmother in Lewisham to keep him liquid, and boasted, somewhat unconvincingly, that even if his father's income was reduced by half, 'it will in no measure affect my happiness'.⁶

Sir John Boyd's ambition can be charted through the growth of the Danson estate up to its sale in 1805 (*see* Figs 2.4 and 7.3). The old house's park comprised about a third of the largely agricultural estate that came together in the late 17th century. Boyd's new park, laid out by Richmond, took in the entirety of this area, some 200 acres (81ha), all of them eventually held freehold. However, he was not content with a mere park: he wanted a working estate, like the best country landowner. In the 1760s, and again in the 1780s, Boyd bought up adjacent farms, amassing about 420 acres (170ha), making his total estate about 620 acres (251ha) in all. Unfortunately, there are no records to show how Boyd managed this productive land, but an estate plan of 1805 identifies, to the east of Danson Road, on an axis with the house, a 'Farm House and Barns, Yards and Buildings, a Pinery, Grapery, Melon Ground and Garden' (*see* Fig 7.3).

In two letters written in 1780 the younger John Boyd estimated the house and land at Danson to be worth about £24,000, but the income from the estate to be a 'mere trifling amount', no more than £500 annually.⁷ The same correspondence reveals that in the early 1770s the family sugar plantation in St Kitt's netted more than £8,000 annually from just 200 acres (81ha), and the Grenadian estate, which was about half this size (and acquired subsequently, through his wife's family), earned him upwards of £4,000 annually.⁸ It is easy to see why the collapse of sugar prices during the second half of the 1770s was such a catastrophic blow to Boyd personally, bringing on a major depression and forcing him to negotiate mortgages on the property and take his son out of the diplomatic service.⁹

In 1780, Boyd senior had to borrow £25,000 from one of his East India Company associates. Before the year was out, however, the loan was called in. Raising even part of the sum owed was, the son wrote, hopeless, since Boyd had 'from gloom dissolved all former connections'.¹⁰ Nevertheless, Boyd did eventually manage to borrow a larger sum from a more reliable source, Captain Nathaniel Smith, who had started out as a clerk in Boyd and Co. Buttressed by a loan of £56,000, Boyd seems by the early 1780s to have begun to contemplate improvements to the house and estate.¹¹ This may have been aided by the sale of the family's shares in the Bance Island Slave Factory off the coast of Sierra Leone.¹² Smith's £56,000 enabled him to spend on a scale unmatched since the early 1760s, extending the park to its present boundaries and buying up more agricultural land on the fringes (*see* Fig 2.4).¹³ It seems likely that the canted bays on the north and south sides of the house were raised in the 1780s. They are first shown in Richard Corbould's view from the south, published in 1794 (*see* Fig 5.2). The alteration involved some risk to the structure as the girders spanning the bays at bedroom-floor level were cut to provide access to the closets added above the bays (*see* Fig 4.8).¹⁴ Raising the side bays also detracted from Taylor's strong axial emphasis (Fig 7.4). The massing of the building was thus made less dynamic, more sedate, and less Taylorian.

Figure 7.4
A reconstruction drawing of the house as it would have appeared after the side bays were first raised in about 1780.

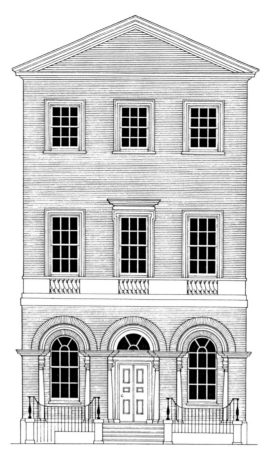

Figure 7.5
33 Upper Brook Street, drawn for the Survey of London.
[MD96/09838, from Survey of London 1977, 121 fig 8b]

Maintaining a gentleman's presence in town

While Boyd might have been keen to involve Chambers in his Danson project, it did not tarnish his allegiance to Robert Taylor, during the 1760s and 1770s, on other projects. As a businessman and gentleman he had to maintain a presence in town and so he employed Taylor on two London town houses.

In 1767, just two years after his father's death, Boyd employed Taylor to remodel an unremarkable house of the 1740s at 33 Upper Brook Street (Fig 7.5). The result, in both elevation and plan, was special, unlike its neighbours. The door of 33 Upper Brook Street was moved to the centre of the three-bay elevation, away from its more usual position beside the party wall, and the rebuilt ground-floor elevation was treated as a round-arched arcade, with a rendered recess set back from the main plane of the façade and relieved by diminutive columns. Above, on the first floor, the window parapets were given an applied, blind balustrade. Spanning the full width of the frontage was a shallow triangular pediment which no longer survives. By these simple, subtle touches Taylor 'Palladianised' the Upper Brook Street house.

Figure 7.6
Grafton Street, looking north. In 1773, Boyd leased no 7, refronted and shown here occupied by Monika Sprüth and Philomene Magers.
[© Richard Lea]

Behind the façade, the stair rises to the left of the vaulted entrance hall that leads directly to an octagonal room behind a canted bay at the back of the house.[15]

Boyd never occupied the property, which, when completed in 1769 or so, he let to Sir Henry Houghton. It seems that he decided to 'trade up' – to swap Upper Brook Street for a more fashionable address in the heart of Mayfair. The new house was 7 Grafton Street, part of an exclusive terrace just east of Berkeley Square, a development that Taylor designed for the Duke of Grafton in the late 1760s (Fig 7.6).[16] Boyd acquired a lease in 1773, but then the house appears to have stood empty until his son occupied the property, for a decade or more, through the late 1780s. At different periods it was let to tenants, before the remaining portion of the lease was sold in 1792.[17] Although the house itself survives, there have been significant changes and it is hard now to appreciate its qualities. Relative to other houses in this fine group by Taylor, Boyd's façade was rather small, but the interiors might well have been exceptional. A number of the surviving Grafton Street properties designed by Taylor are buildings of the highest quality.[18]

John Boyd II at Danson, 1800–1805

After Sir John Boyd's death in 1800, it was John Boyd II who realised his father's plans for the estate by spending large sums between 1801 and 1804.[19] Although as a young man John Boyd II confessed to hating the house, associating it with his father's depression, the family's financial trouble and the ruination of his own budding diplomatic career, he now set about leaving his own mark on the property.[20] However, by 1805, debts that had accrued over the previous 20 years proved too much and he sold the house and its estate.[21]

Some of the works that John Boyd II carried out were minor, and we cannot be sure in every case that it was he, rather than the next owner, John Johnston, who was responsible for them. Of several rooms that were redecorated in 1801–4, the library received the most attention. A cream painted frieze of urns and arabesques, executed in composition on a mahogany board, was applied to the wall immediately below the cornice, and the large round plaster plaque from above the fireplace was removed and

Figure 7.7
Coade stone chimney pots on the roof, date-stamped 1801, one of several alterations that John Boyd II made to the house after his father's death in 1800.
[B880014/50]

carefully reused as an overdoor in the closet next to the principal staircase. It was framed in its new location by arabesque panels that match the library frieze. The large room above the entrance hall was divided in two, and the original door in the centre of the south wall was blocked. Two new doors were formed, cutting through the trussed partition. The house was also given a set of new chimney pots (Fig 7.7).

Unquestionably, the most dramatic work was the demolition of the wings and their linking walls and the building of a new stable block with reused stone and other materials from the earlier structures (Fig 7.8 and 7.9 and *see* Figs 9.4 and 9.10).

Certainly the planning of country houses had moved on since the 1760s. Humphry Repton's *Observations on the Theory of Landscape Gardening*, 1803, advocated placing stables away from the house but not remote from it, so guests could see but not smell or hear them.[22] The Danson stables were positioned in just this way: grand and imposing in aspect, when seen from the drive, but distant from the pleasure grounds around the house. A small plantation, formed around the stable when it was new, screened it further from the main house.[23]

Unfortunately, the removal of virtually all the fittings from the stables makes it difficult to say exactly how the different parts of the building were used. The sale catalogue of 1805 noted accommodation for 3 coaches and 16 horses, with a kennel for pointers.[24] Traces of the stalls for 4 horses survived in the southern half of each of the 2 wings. The wide openings

*Figure 7.8
Reconstruction of Danson soon after John Boyd II sold the property to John Johnston in 1806. In 1801–4, Boyd II had dismantled the quadrant walls and service wings, reusing the material to build the present stable block, north-west of the main house, and built single-storey rooms in the areas either side of the entrance. John Johnston built a new orangery to the west soon after he acquired Danson (see Fig 8.2).*

in the central range were for coaches. Heated rooms to either side of this could have served as offices or a groom's accommodation, with the spaces to the outer corners of the north range serving as the stable areas themselves. The first floor of the centre range had heated chambers at either end, again indicating offices. Haylofts were probably fitted into the first-floor areas of the wings above the stalls but these areas were later lined out as office accommodation.

Although pieced together from reused architectural features, the stable block is a well-considered, handsome building, elegant and monumental. It suggests the hand of an architect. We know it was built between 1802 and 1804 but not by whom. There is, however, direct documentary evidence linking the estate to the eminent City architect George Dance the Younger (1741–1825). Dance became involved with Danson in 1794 in connection with a business bond for £32,000 which Boyd and his partner John Trevanion had taken from Captain Nathaniel Smith and his wife Hester, née Dance (the architect's sister). When Captain Smith died, Dance was appointed one of the trustees overseeing this bond and was named in a lease of August 1800 which transferred the obligation of that bond to John Boyd II.[25] Dance appears again in a lease of 1802, this time selling one acre of land in Welling owned with his sister to Boyd, and finally is named in the leases of July 1806 which gave the new owner, John Johnston, the freehold to the property.[26] The documents of 1806 refer to £1,000 transferred to Dance personally on behalf of the trustees, though in consideration of what exactly is not said. Part of this sum perhaps represented payment for Dance's role in the rebuilding of the stable block.

Dance is normally associated with schemes of urban development and improvement, and his main body of work was for the Corporation of London. In the latter part of his career, however, he did take on a number of small-scale commissions for modernising country houses. Around 1800 he remodelled Dorton House for the Marquis of Camden. In 1803 and 1804, he rebuilt East Stratton and Coleorton Hall. In 1805, Dance started alterations to a stable block at Laxton Hall in Northamptonshire. Finally, in 1807, he was commissioned to remodel Camden Place in Chislehurst, not far from Danson, for the banker Thomas Bonnar, work that included a new stable block (now demolished).[27]

Boyd is unlikely to have undertaken such considerable works without some hope of staying at Danson for years to come, which begs the question of where these funds came from. As recently as 1794, the elder Boyd had taken a mortgage bond on the property for some £40,000, a bond still outstanding when the estate was settled on his son. The evidence of the younger Boyd's bank account suggests he could not possibly have cleared it, and so it is all the more surprising that he pressed ahead.[28] The sale of his father's pictures in two lots in 1802 and 1804 would have helped.

In the event, by 1805, the Boyds' means were exhausted, and on 27 June that year the auctioneers Coxe, Burrell and Foster came in to sell Danson.[29] The estate plan prepared at the time of sale shows Boyd II's recent improvements (*see Fig 7.3*).[30] The wings have gone, and plantation nearly encloses the stable block, effectively concealing it from the house. The total freehold property amounted to 563 acres (228ha) with another 56 acres (23ha) on leasehold.

Figure 7.9
The stable yard in the late 19th century, looking north-east.
[© Bexley Heritage Trust. (B960788)]

Apart from the stable accommodation noted above, there was the substantial Home Farm or 'Bailiff's Residence' as it was called in the eastern part of the park. It provided stabling for 12 horses, a dovecote, granary, cow pens, piggery, and an 'extensive brick-built barn'. Attached were hothouses, a grape arbour and a yard containing a large quantity of 'ready Squared' stone, left over, undoubtedly, from the earlier wings and quadrant walls.

The highest bidder at the auction was John Johnston, a merchant of Newman Street in the parish of Marylebone.[31] Johnston paid £36,000, of which £1000 was for books, artworks and furnishings, and as compensation to Boyd for timber on the estate and for a recently constructed lime kiln.[32] Johnston took possession in 1806, while Boyd and his young family moved to a town house in Baker Street.[33]

In his negotiations with Johnston over the value of timber and other commodities, Boyd made a passing reference to the new owner 'carrying on the work at Danson'. On 9 September 1806 Boyd remarked, after a visit to Johnston's City solicitors, 'I hope to be able to give you the Information you require, together with some Points Explanatory of the Vase, Drawings of intended additions to Danson'.

It continues:

> There are at Mr Coxe's in Pall Mall two large Views of Danson, which were painted before I made alterations there … . I ordered them to be bought in when my Pictures were sold by Coxe, thinking the Purchasers of Danson might be glad to have them; If you think proper to call & look at them, you may see them at any time, & if it is your wish to have them, you may do so upon very easy terms – Should you wish to have them [the paintings] altered according to the present state of the place, I can recommend Barrett to you, who lives in [Baker] Street near me, and is Brother to him who painted them for my Father; he was a pupil to and brought up under his Brother.[34]

It is this letter, taken together with the sale catalogues published for the auction of Boyd's picture collection, that establish the attribution of the painting of the north view of Danson to George Barrett, but they also show something of the younger Boyd's concern that Johnston should continue his work there, and the role that country house pictures could play in the life of an estate, both as a reminder and an active record.

8

Nineteenth-century Danson

John Johnston, 1806–28 and Hugh Johnston, 1828–63

John Johnston (1745–1828), the new owner of Danson in 1806, had much in common with its builder (Fig 8.1).[1] Like the Boyds, the Johnstons were of Scots lowland stock and had made their way to London via Northern Ireland in the late 17th or early 18th century. Although the details of the family business are still hazy, John is known to have been a merchant in the City of London with offices in Coleman Street Buildings. And, like Boyd, he was involved with the West Indies (Trinidad and Tobago).

Johnston first took Danson as a suburban retreat when his business was expanding, and then later decided to retire there. He was involved in the affairs of the parish church of St Mary the Virgin, Bexley, serving as a Poor Law Guardian (1810 and 1821), and a Churchwarden (1813, 1814 and 1823). A fine

Figure 8.1
A portrait of John Johnston (1745–1828), probably painted in the late 1790s. He occupied Danson from 1806 until his death.
[© Bexley Local Studies and Archive Centre. (B960529)]

plaque to his memory survives in the church. He contributed to the building of a Chapel of Ease in Oaklands Road, the predecessor of Christ Church, Bexleyheath, and gave liberally to the construction of the parish school.[2]

Johnston did very little to Danson House, not surprisingly since it had just been repaired and renovated, but he did carry out extensive works on the park. These are recorded on a poorly preserved estate plan which may have been drawn up in connection with Johnston's will, proved on 21 March 1829.[3] He closed off the open view of the main entrance of the house from the London–Dover Road, to the north (Welling High Street) with a plantation, intensified the planting at the western extreme of the lake, and infilled clumps of trees to the south of it, creating a firmer border, but leaving the great expanse from the south front of the house down to the water. The effect of these works would have been to create a private enclosed area, a transformation reflecting emerging 19th-century tastes. The same plan also shows the orangery, the appearance of which is known from photographs of about 1900 (Fig 8.2). Built about 100 yards west of the house on a simple rectangular plan, and later known as the 'Winter Garden', it incorporated eight rusticated columns from the arches in the quadrant walls in a five-bay temple colonnade. It appears on the Ordnance Survey map in 1936 but the date of its demolition is not known. It may have been removed when the Doric Temple was transferred to St Paul's Walden Bury in 1961.[4] Since the orangery is not shown on the estate plan of 1805, it is likely that Johnston built it soon after he purchased Danson in the following year, making use of the 'ready Squared' stone then stored at Home Farm.

Johnston also commissioned designs for a pair of gate lodges and gates from the architect George Repton, a son of Humphry Repton and an assistant in the office of John Nash, although

Figure 8.2
John Johnston's orangery, photographed in about 1900. It was built to the west of the main house soon after 1806, using rusticated stone columns from the Doric arches in the former quadrant walls (see Fig 0.4) and demolished between about 1950 and 1961.
[© Bexley Local Studies and Archive Centre. (B960768)]

neither appears to have been built in the manner shown in these drawings (Fig 8.3).[5] It was probably Repton, however, who designed the Gothick-style rustic gamekeeper's lodge that stood near the west entrance to the present park from Danson Lane in the *cottage orné* manner popularised by Nash and the elder Repton in the 1810s and 1820s. It was demolished after 1948 (Fig 8.4).[6]

The terms of Johnston's will explain in part why the main house survives with so few significant alterations. The bulk of the estate, £38,000, was to be invested in government securities, with most of the income settled on his wife, who was given liberty to reside in the house with Hugh for seven years. He was given £14,000 outright in consideration of this. A codicil executed one month before his death, however, gave Johnston's wife the house and estate outright until her death, in addition to all the goods within it, including the family furniture and library, but only on condition she reside there permanently and continue to run the park and farm. Hugh, the eldest surviving

Figure 8.3
A gate lodge at Danson Park, designed by George Repton in the 1810s or early 1820s. It is probably the lodge that stood west of 'Little Danson' (see Fig 8.5), at the junction of modern-day Hook Lane and Bellegrove Road. It was demolished in the early 20th century.
[RIBA Library Drawings Collection SD111/6]

Figure 8.4
A 'Gothick'-style rustic gamekeeper's lodge built for John Johnston, probably to designs by George Repton in the 1810s or early 1820s. It stood at the west entrance to the park, facing modern-day Danson Lane, and was demolished after 1948. Late 19th-century photograph.
[© Bexley Local Studies and Archive Centre. (B960767)]

son, continued the family business (he had been made a partner in 1818) and stayed in Bexley, building Little Danson (Fig 8.5). It was constructed to the designs of George Stanley Repton (1786–1858), as a substantial but modest villa north-west of the main house on the edge of the park (it has since been demolished).

The parish tithe assessment made at about the same time, in 1840, shows that the estate was still a reasonably substantial agricultural concern, with a large number of plantations – established by the Boyds – and a lesser amount of pasture. Small parcels of land were let to farmers, and there was a good-sized cherry orchard of about 10 acres (4ha).[7] Hugh and his family stayed on in Little Danson until his mother's death in 1860, after which the family enjoyed a brief spell in the main house until its sale in 1863. It was during this period that Hugh's daughter Sarah painted the five watercolour views of the building, inside and out, which now constitute our most important visual record of the house in the Victorian period (Fig 8.6). They then decided to move to a more modest property, without an agricultural estate – Iridge Place at Hurst Green, in Sussex.[8]

The Johnstons took with them many of the pieces that had been purchased from the Boyds

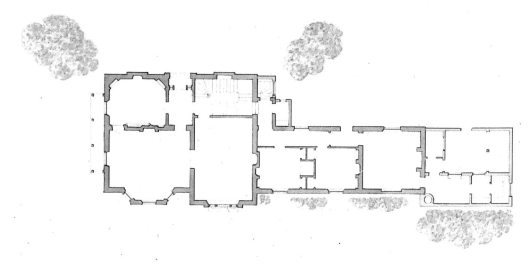

Figure 8.5
George Repton's perspective of 'Little Danson', built for John Johnston's son, Hugh, and his family, in the village of Welling in the 1810s or early 1820s, south of the main road. The plan includes an addition to the three-window range at the east end, in a lighter wash, which was built to a different plan. The perspective shows the south side of the house, looking towards Danson Park.
[RIBA Library Drawings Collection SD114/3]

Figure 8.6
Watercolour by Sarah Johnston of the house from the north. It can be dated to 1860 or soon after, as it shows a hatchment hung in the pediment in memory of the artist's grandmother, Anna Johnston, who died in that year.
[© Bexley Heritage Trust]

in 1806, including a set of four bronze figurines copies of well-known Roman sculpture, such as the Borghese Gladiator, and a carved head of a river god in stone, cut to receive a water spout in its mouth. They also transported the enormous Piranesi Vase to their new home, but soon decided to sell it. Hugh offered the piece to the British Museum, and the Trustees installed it in a temporary shelter under the great south portico while negotiating a price. However, in July 1863 the Museum flatly refused to pay the £500 Johnston was asking, as they regarded the vase as an assemblage of unrelated antique fragments and were unhappy about the degree of restoration by Piranesi (ironically, part of its interest today). They did continue to display it in the south portico until, in 1868, Johnston agreed to sell it for just £100.[9] As for the estate in Bexley, this had been sold some years before, on 13 June 1863, in 10 lots by the auctioneers Norton, Hoggart and Trist. A rare early photograph records the east side of the house at the time of the sale (Fig 8.7).

Figure 8.7
Photograph of the house from the east in 1863, the year it was sold to Alfred Bean.

Alfred Bean, 1863–90

Danson's new owner was Alfred Bean (1823–90), a railway engineer who had trained in his father's engineering and contracting firm, Bean and Jackson (Fig 8.8). The company had established itself through canal building, and had a local business interest, first the construction of the North Kent Railway and then, in 1866, the Sidcup line. Later, in 1883, Bean and Jackson formed a company to build a rail link from Dartford to London via Welling, a line that drove the suburbanisation of this area. Bean was the company chairman.[10]

Sarah Johnston's watercolours of about 1860 show that the house had changed little since it was first built and yet, even though Bean was to alter it far more significantly than his predecessors, his work cannot be described as substantial. Instead, it can be seen as modernisation in line with standards of Victorian propriety and comfort (Figs 8.9, 8.10, 8.11, 8.12 and 8.13). While some of the changes reflect personal interests in engineering, others express more general concerns with household efficiency and hygiene. His ready access to the stuff of engineering is perhaps most clearly demonstrated in the unusual quantity of iron he built into the fabric. On the basement floor,

Figure 8.8
Portrait miniature of Alfred Bean (1823–90) in about 1880. He bought Danson in 1863, having made his fortune developing the North Kent Railway.
[© Bexley Heritage Trust. (F960612)]

timber sash windows were replaced with a cast-iron glazing system more commonly used in warehouse and factory buildings, and timber lintels were replaced with heavy 'I-section' iron girders. Failing trussed floor girders above the saloon were repaired with large steel flitch plates let in between the two halves of each of the original girders.

Figure 8.9
The dining room in the sale catalogue of 1922. The gasolier and plate-glass mirror in the round-arched recess date from the 1860s.
[© Bexley Heritage Trust. (B960771)]

NINETEENTH-CENTURY DANSON

Figure 8.10
The saloon in the sale catalogue of 1922, and not much changed from Alfred Bean's refurbishment in the 1860s.
[© Bexley Heritage Trust. (B960779)]

Figure 8.11
The library in the sale catalogue of 1922. Alfred Bean's modifications in the 1860s included glazed doors to the bookcases, window pelmets and a shallow frieze beneath the cornice. George Barrett's painting replaced the circular plaque over the fireplace. The elaborate gasolier was probably added in about 1880.
[© Bexley Heritage Trust. (B960775)]

Figure 8.12
The eastern half of the principal bedroom on the south side of the house in about 1920. Alfred Bean inserted a partition wall in the late 19th century, to create a study on this side of the original bow-shaped room.
[© Bexley Heritage Trust. (B960781)]

Bean updated the heating arrangements with large hot-water or steam-heated radiators in the entrance hall concealed by slate tops and decorative cast-iron grilles. With the exception of those on the principal floor, all the fireplaces were furbished with cast-iron grates. In the principal bedroom, he installed a substantial new Italian marble chimney piece in place of the original, which was relegated to the north wing. As well as making the house warmer, Bean also made it more convenient, with a system of bell-pulls and improved plumbing, the latter involving a new slate cistern above the stairs to the attic, with a capacity of about 935 gallons (4,250 litres). This probably served a bathroom and flushing toilets in the closets above the side bays. Bean also addressed lighting levels inside the house when he replaced almost all of the original box window shutters on the bedroom and ground floors with sliding vertical sash shutters and splayed reveals, and replaced the oil lamps shown in Sarah Johnston's watercolours with gasoliers.

Not surprisingly, Bean's alterations reflect developing technologies. Developments in the production of plate glass in the 1840s had made large sheets much more affordable and glass therefore features prominently in the modernisation of the house. The original mirrors in the dining room and saloon were replaced with larger mirrors in a fashionable French rococo taste (see Fig 8.9). In the library, the original bookcases were given glazed doors, some mirrored, and others incorporating pleated red satin. New doors installed in the entrance passage on the ground floor were without glazing bars and many of the original glazing bars were removed from the ground-floor windows to accommodate larger panes. Similarly, the door furniture was updated with hardware bearing the names and references to patents of manufacturers who led the field, Hobbs & Co and Chas Collinge & Co. These stamps help us to date the refurbishment to around 1865, immediately after Bean purchased the property.

The organisation of Bean's household was reflected in the subdivision of some of the rooms and changes to access routes through the house. A staircase was built at the north end of the kitchen and a flight of steps was cut through the cellars under the perron to improve communication between the ground and basement floors. The kitchen itself was opened up by the removal of the cross-walls, and the present butler's pantry (now the shop) was created in the middle of the ground-floor entrance hall. On the principal floor, a door was cut through, linking the saloon and the closet next to the stairs, and making the closet a corridor between the entrance hall and the saloon. The main rooms could thus be used more individually, not only as a suite. On the bedroom floor, the principal bedroom was subdivided down the middle, though this alteration was probably later in Bean's tenure (*see* Fig 8.12). The reorganisation of the layout resulted in shorter access routes, but also greater compartmentalisation of domestic functions.

Not all Bean's changes were pragmatic; some show the influence of fashion, and this is most evident in the interiors of the principal floor. Taylor's Palladian interiors were given a High Victorian polish with parquet floors installed in the dining room and library; glazed and mirrored doors in the library bookcases; heavy sideboards with naturalistic carving in the dining room, and new plate-glass mirrors, neo-rococo window pelmets and new damask in the saloon. All this was supplemented by a dense scheme of furnishing, to create an interior that was quintessentially of the late 19th century. In line with developing paint technology, Bean's paint schemes were more colourful than those they replaced. In the dining room, the joinery was picked out with cream and pink oil paints, and the cornice with blue distemper. In the library, details in the cornice were picked out in yellow and emerald green, while in the stairwell, the columns were marbled and the stair balusters painted bronze-green.

Bean's wife Ann, born in Staffordshire in 1823, survived her husband by 30 years, and for the second time the house was untouched for a significant period of time because a wife outlived her husband. By the time of the 1901 census, Mrs Bean was no longer living at Danson: instead the census entries paint a picture of an estate managed by a skeleton staff consisting of a caretaker for the house, a gamekeeper, a gardener, a coachman and a groom.[11] Then in 1921, after the Great War, when the taste for great houses and Palladian architecture had ebbed and Alfred Bean's efficient family mansion had become obsolete, his widow died aged 97. The pressure for more intensive suburban development was increasing, and in 1922 and 1923 about 400 acres or so of farmland (162ha) that Boyd had added to Danson were sold for residential development, leaving only the park and house.

Figure 8.13
The breakfast room in about 1920.
[© Bexley Heritage Trust. (B960780)]

9

Danson House and Park since 1924: decline and restoration

When Danson House and Park first came up for auction in 1922 there were no immediate takers. Grand classical houses were unpopular between the wars, especially those located in the metropolitan 'fringe' areas then being developed, with semi-detached estates and shopping parades.[1] While councils were keen to convert such private grounds into public parks and sports fields, the houses that came with the open space were considered liabilities or, in some cases, despised as symbols of traditional privilege, contrary to modern, democratic values.[2] When it came up for auction again in 1924, the Urban District Council of Bexley stepped in.[3] The council's motivation was to secure the open space as an amenity for the residents of the freehold suburban housing being built in the locality at an astonishing rate. 'A magnificent defence against the Brick and Mortar Brigades from London' was how the council described Danson in *The Bexleyheath and District Handbook* for 1929, believing that the historic mansion and park gave Bexley a competitive edge over neighbouring authorities.[4]

Danson was opened with grand ceremony by Princess Mary on 13 April 1925 (Fig 9.1).[5] Thereafter, public viewing of the principal rooms took place on Wednesday, Saturday and Sunday afternoons. The rooms could be hired for wedding receptions, tea dances and whist drives. The most popular attraction was a tea

Figure 9.1
Princess Mary presided over Danson's opening to the public, in a ceremony held on 13 April 1925. A sundial in the south lawn commemorates the occasion. [By permission of Bexley Local Studies and Archive Centre. (DP094082)]

Figure 9.2
The house from the south-west in June 1953, with decorations for Her Majesty Queen Elizabeth II's coronation. The former breakfast room on the ground or rustic floor was a tea room, with tables and chairs placed out on the south lawn in fine weather.
[By permission of Bexley Local Studies and Archive Centre. (DP094081)]

room created in the former breakfast room on the ground floor.[6] On the first floor was a local museum with displays on the area's natural history and archaeology (Figs 9.2 and 9.3).

Despite the house's popularity, however, Bexley Urban District Council, which became a Borough Council in 1937, spent little on maintenance.[7] When the fine plaster ceiling in the saloon began to bulge, the solution was to rope off the area underneath.[8] During the war, when the Council was prohibited from spending money on maintenance, the Air Raid Precautions Department commandeered the house, and the kitchen was home to the Civil Defence Headquarters. After 1945 visitor numbers declined, and essential repairs, put off for 20 years, were done poorly. In 1951, the pediments and entablatures to the windows on the *piano nobile* were cut back and rendered over.[9]

In 1953 Danson was listed grade II as a building of special architectural and historic interest, but this made little difference to the way it was maintained by the local authority.[10] The early 1960s, however, saw a brief revival of interest in the house, following the restoration of the organ in 1957–9.[11] On 3 December 1959 there was a special concert to celebrate the instrument's return. More concerts followed, and in July 1963 a recital of period music was broadcast live from Danson on BBC radio.

The Georgian buildings on the estate fared worse. Chambers' little temple on the north side of the lake was also listed in 1953 but became dangerous, and in 1961 found its way to the grounds of a house designed by Taylor's contemporary, the architect James Paine, St Paul's Walden Bury near St Albans, where it survives today.[12] The early 19th-century stable block only narrowly escaped total

Figure 9.3
The local history museum in the former north-west bedroom on the first floor (1930s–50s).
[By permission of Bexley Local Studies and Archive Centre. (DP094083)]

Figure 9.4
The stable block of c 1805, as it appeared in 1993.
[M960098]

Figure 9.5
Danson from the south-west in 1993. The upper part of the canted bay on the west side of the house had been demolished, leaving a gaping hole into the bedroom floor.
[M960097]

demolition in 1964 after the Borough Engineer concluded it was beyond feasible repair (Fig 9.4).[13] In 1969, when another proposal for the demolition of the stable block was submitted, local residents joined forces with the established amenity and historical societies to complain about the degradation of the whole site. The Minister of Housing and Local Government decided to call in the application for demolition of the stables for determination at a public inquiry, which met on 14 April 1970. The decision went against the Borough.[14]

Meanwhile, the condition of the house was a source of growing concern. It was closed to the public in 1970, and the organ and dining-room paintings were removed the following year. The repair costs, the listed status and location in a public park were serious constraints on any private use, and the grants on offer did not offset the site's negative commercial value. Finally, in February 1985, Bexley accepted an offer from a firm specialising in decorative plasterwork, to restore the house as a private residence using their own craftsmen. The interiors would be a showcase of the firm's work: the stables could be converted to craft workshops for the business.[15] However, in December 1990, after delays caused by storm damage in 1987, and disputes over the level of grant available for the repair of the building, the owner, Mr Laurie Taylor, informed Bexley that he would be quitting his lease and leaving the property. In doing so, he observed that the marble chimney pieces on the principal floor were easy to remove, and the local authority should arrange for security after his departure. On 17 February 1991 he left the property for his house in Antigua, and on the following day council officers discovered that every fixture had been removed, not just the three marble chimney pieces designed by Chambers, but decorative plaster fittings, doors, and even doorcases.[16] Taylor informed the authorities that the items were in safe storage, though he did not say where. In June 1992, HM Customs retrieved the three chimney pieces. The rest of the fixtures were located in a container in Dagenham.[17]

The property had now reverted to Bexley, which was seeking a development partner for the restoration of the building. Negotiations with one company collapsed, its managing director citing the poor economic climate, high interest rates and falling property values.[18] English Heritage now started to explore the possibility of restoring the building itself and then selling it on to a private user. Negotiations with Bexley were slow, the principal sticking point being the national conservation body's demand for a 999-year lease. However, in May 1994 the architects Purcell Miller Tritton were appointed, alongside Tapley and Eadworthy as the quantity surveyors. A scheme design was commenced at once, and English Heritage submitted a proposal to the Department of National Heritage, who required a written assurance that the organisation was not contemplating taking the house permanently into care. Finally, in March 1995, the Secretary of State for Heritage, Lord Astor, met Sir Jocelyn Stevens, then Chairman of English Heritage, to deliver formal approval.

When English Heritage took on Danson in 1995, everyone's vision was shaped by the suite of photographs taken in 1922 for the sale of the estate, which show, not a Georgian interior, but a glossy, High-Victorian one (*see* Figs 8.9, 8.10 and 8.11). It was assumed that the decorative interior of the house would be conserved to match these images. On the other hand, the furnishings were gone, and the finishes were beyond the point of respectable decay, challenging the 'conserve as found' option (Fig 9.5). There was obviously considerable research to be done before the exact treatment of the historic interiors could be decided on. Paradoxically, this was enabled by the poor state of the building which presented an opportunity to analyse how it had first been put together. The research revealed that despite the superficial Victorian appearance of the interiors, these later owners had actually removed very little of the original fabric.

As far as the exterior of the building was concerned, the first task of the architects, Purcell Miller Tritton, was to survey the property and draw up a costed repair schedule accompanied by a complete photographic survey.[19] These images show the exterior in a dreadful state, but much of the decay turned out to be superficial (Fig 9.6). The roof structure was largely intact, but the masonry was very wet, which in turn threatened the timbers bedded in it.

Robert Taylor had originally intended a smooth stone finish for the exterior of the building, that would shed water discharged by the roof eaves, but be porous enough to permit quick drying. As this stone surface was no longer smooth, a decision was made early on to limewash the whole of the exterior. This protective surface, or 'shelter coat', weathers down over time and requires periodic renewal.[20] First, the degraded stone surfaces were cut back to the solid and built up again with lime mortar on stainless-steel armatures fixed to the remaining stone (Figs 9.7 and 9.8). The projecting window heads of the ground floor had to be replaced in new stone. Historic photographs showed that they had suffered most in the repairs of 1951 when they were cut back flush with the wall face and then re-dressed with mouldings, crudely executed in cement render. Fortunately, just enough of each moulding profile survived to allow full reconstruction

Figure 9.6
The south elevation, from the south-east in 1993. The removal of paint from the 1960s had exposed a patchwork of unsympathetic stucco repairs and decaying stonework.
[F931064]

Figure 9.7 (above)
Repairs to the stonework of the principal floor in the north-west angle in 1997, using lime mortar, applied slowly, in at least three layers.
[F970472]

Figure 9.8 (above right)
Repairs to the rusticated stone in the north-west angle in 1997. When the flat-roofed additions were built in 1801–4, the original rusticated stone was cut back. The chamfered blocks were restored in lime mortar, except in the area around the window, which was executed in new stone.
[K970388]

Figure 9.9 (right)
The head of the pedimented window on the south wall of the dining room in 1996, after the removal of the cement render. The application of this render in 1951 had involved hacking away the original cornices.
[F90149]

(Fig 9.9). The cornice above the main entrance door had suffered at the same time; although most of the detail survived, the modillion brackets, which were entirely missing, had to be recarved.

By the summer of 1996, the experts' understanding of the historic interiors of Danson had deepened to the point where they were completely reappraising their original approach. It had become apparent that the Victorian finishes were relatively superficial, and that there was more than enough evidence remaining of the Georgian interiors of the principal rooms to consider reconstructing them. In the light of the ease with which the Georgian decoration schemes could be read and appreciated, the case for simply conserving the existing decoration seemed to make little sense. Why invest in conserving a second-rate Victorian interior, restoring large plate-glass mirrors and sideboards, glass-fronted bookcase doors and gasoliers to furnish the shell of a magnificent Georgian villa? The few gaps that might be left in the team's knowledge of the Georgian interiors could be filled in by copying the surrounding decoration or by referring to other Taylor works.

With the new approach in mind, building analysis, paint research and dendrochronology (dating timber through the analysis of tree-ring patterns) continued apace to fill out understanding of the building's development. The extent to which these separate disciplines were integrated makes Danson unusual. All too often, paint samples are taken with the sole purpose of determining the earliest decorative schemes and yet the stratigraphic sequences of paint observable in each sample have much more potential. At Danson, paint sampling enabled the identification of a succession of decorative schemes for each interior. Contemporary but physically unconnected features on opposite sides of rooms could thus be linked through their historic paintwork.

The ultimate shift in the balance from conservation to restoration was decided by the chance discovery of a suite of amateur watercolours in the private collection of a descendant of Hugh Johnston, one of the 19th-century owners. These showed the dining room and saloon as they appeared in the early 1860s, before the Victorian modernisation (*see* Figs 6.13, 6.14, 6.32, 6.56 and 8.6). Here was a wealth of extra evidence to be used in the re-creation of the 18th-century interiors. Fortunately the watercolours did not include details of the building's exterior, as the stonework, roof and structural timber repairs were already underway. These were able to go ahead and were largely completed by November 1996, enabling the scaffolding to be taken down over that winter.

The second phase of the repair and restoration work, partly enabled by the sale of the stable block to Bass Taverns Ltd, covering work to the principal interiors, began in September 1996, and was completed just over a year later, when excitement for the project was running very high. By this time, the chimney pieces in the principal rooms and the doors had been returned and most of the paintwork completed. In the library, the plaster roundel was restored to its original position above the fireplace (*see* Fig 6.46). Works to the stable block progressed at great pace and in September 1997 Bass Taverns opened the rehabilitated stables for business as a pub/restaurant (Fig 9.10). Soon afterwards, work on the main house stopped until a suitable occupier could be found. It was not until early 2002 that negotiations with Bexley Council and Bexley Heritage Trust were sufficiently advanced to allow works to proceed. The decision was made on the basis that Danson would be managed by the charitable trust as a historic house open to the public. This third phase of English Heritage works saw the fitting-out of the interiors in a way that would satisfy this requirement.

It was on the principal floor of the house that John Boyd was best able to demonstrate his wealth, taste and knowledge of the arts to his visitors. With views over his landscaped park, each of the four principal rooms amounted to an essay on the classical idyll. The conservation of the rooms on this floor accordingly went beyond repair to present the surviving fabric in such a way that this aesthetic could be appreciated again. Victorian features that would have obscured this 18th-century vision were not reinstated, but simply put into store.[21]

In the entrance hall, Victorian radiators and sculpture pedestals were removed, and the room returned to its original colour scheme (Fig 9.11). The most striking change in the

Figure 9.10
The stable block in 2009. The sale of the stables to Bass Taverns, in 1996 helped fund some of the works to the main house.
[DP076962]

Figure 9.11
The entrance hall in 2009. The original door was made with an unusually large pane of glass. The Ionic door surround matches that to the staircase on the opposite side (see Fig 6.10).
[DP076245]

dining room was the removal of the large 1860s mirrors and sideboards from each end of the room (see Fig 8.9). Sarah Johnston's two watercolours show the original Georgian mirrors still in place in 1860, and their existence was confirmed by their outlines in layers of paint applied to the plaster behind the Victorian mirrors (see Fig 6.15). The watercolours also enabled the identification of a set of gilded fragments in the Bexley local studies collection. With these discoveries, it was decided to proceed with the reconstruction of the original mirrors.[22]

In the saloon, the room's original drama was restored by lining the walls with blue Chinoiserie wallpaper in 2003; re-creating the mirrors to match those in the Sarah Johnston watercolour; restoring the ceiling arabesques removed in the 1930s; cleaning the original gilding and repainting the walls (Fig 9.12).[23]

The sobriety of the library was re-established by the reinstatement of the large plaster plaque above the fireplace and the bookcases without their Victorian glass fronts, as well as the restoration of the southern bookcase – its Victorian block pediment removed – and the painting of the walls with green verdigris.[24] The organ, which had been removed to Hall Place for safe-keeping was overhauled, reinstated and retuned to a baroque temperament (see Fig 6.46). In the stairwell and landing, the service stair was reactivated by the restoration of its wrought-iron balustrade and later paint was removed from the panelled dome to expose its original trompe l'oeil painting.

In the autumn of 2003, English Heritage returned the lease of Danson to the London Borough of Bexley who assigned it to the Bexley Heritage Trust in February 2004. The Trust is a non-profit organisation set up in 2000 to manage Hall Place and provide a focus for heritage in Bexley; it also manages the Bexley Museum Service. The first task for the Trust, in conjunction with Bexley Council was to furnish the interior of the house. This included the return of the Barrett painting to its position above the saloon door to the library, the replication of Vernet's painting above the fireplace, and the re-creation of the picture frames according to William Chambers' designs. The Trust also set about reworking the setting of the house. The principal improvement in this area has been the removal of the beech hedge and the restoration of the original ha-ha. Now, the house is once again seen to be the focus of the landscape, with no visible separation from its surrounding park.

The house was reopened in March 2005 under the management of Bexley Heritage Trust. As a Georgian villa open to the public, Danson redresses an imbalance between east and west London. In the 18th century there were as many villas in parks in the east as there were in the west, and they were equally important for shaping the growth of suburban London. Danson now serves as a prime example of a Palladian villa, and its visitors can appreciate something of its original function and aesthetic. While promoting this valuable resource, the Trust aims to secure its maintenance with extra income generated by opening the house for weddings, functions, and concerts.

The Danson project has been well received. In 2004 it received a Georgian Group Architectural Award for a conservation and restoration project showing vision and commitment in the restoration of a Georgian building and landscape. In 2005 this endorsement was echoed by the Royal Institute of British Architects (RIBA) with the award of a Conservation Commendation in recognition of best practice in the conservation field.[25] On 26 July 2005 Danson House was formally opened by Her Majesty the Queen.

Figure 9.12
The north wall of the saloon in 2009.
[DP076983]

APPENDIX 1

Transcription of Revd Joseph Spence's comments, May 1763

Joseph Spence's notes on the landscape at Danson are in the Beinecke Rare Book Library, Yale University, New Haven, USA: Osborn Shelves, Spence Papers, MSS 4/6/171. They consist of four sheets of folded paper without numbers. Some of the text has been published in R H King, 'Joseph Spence of Byfleet', *Garden History. The Journal of the Garden History Society*, **8** (no 3 winter 1980), 77–114, at 87. Further notes on Spence were published by King in the same journal, **6** (no 3 winter 1978), 38–64, **7** (no 3 winter 1979), 29–48, and **8** (no 2 summer 1980), 44–65.

For ease of reference here, the four sheets of notes are assigned folio numbers. Folio 1 appears to have been used as a cover for the other folios, since the recto consists simply of a title and date. While the notes and plan on the verso are concerned with the water supply, it is not clear how they relate to the estate and the house (Fig App1.1). The notes on Folio 2 describe the views from the north and east fronts of the house at the centre of the park: verso, they describe the views from the south and west fronts. There is no reference to the woods and fields named on the plan, folio 4,

Figure App1.1
Revd Joseph Spence. Sketch plan, in pen and brown ink and pencil, outlining the arrangements for the water supply to the house. North is to the right of the page. [Joseph Spence Papers. James Marshall and Marie-Louise Osborn Collection, Beinecke Rare Book and Manuscript Library, Yale University]

which suggests that either it was not to hand or that it had not been executed before the notes were made. The notes on Folio 3 set out how the views described on Folio 2 could be improved through landscaping. The comment on the view west refers to 'Mr Richmond's Plan', suggesting that it might be too blocked by his proposed planting. Folio 4 consists of a plan of the estate showing the plots of land, fields, and woods of which it was composed, together with the buildings, lanes, and waterworks before the landscaping in Nathaniel Richmond's plan was put into effect. It is an accurate measured survey drawing, or a copy of a measured survey drawn at a scale of 1:2500, although this is not specified. This scale is deduced from the fit of the drawing with the Ordnance Survey surveys of the site in the 19th and 20th centuries. The table on the right of the drawing records the area of each of the numbered plots that made up the estate and park in acres, rods and poles. The notes and the main plan are executed in ink. The small plan on Folio 1, verso, is in pencil, and there are pencil annotations on the notes. The notes in ink record what Spence saw in the landscape from each front of the house. Since the pencil notes function as annotation to the ink notes, they are clearly secondary. However, since they represent further observations on the views, it would appear that at least some of the pencil annotations were also made on site. This suggests that there is little reason to doubt the date of the visit, 'May 63?' and the location 'at Danson', even though they are pencil additions to the title, 'Mr Boyd's', written in ink on folio 1, recto.

The following conventions have been adopted in the transcription which follows: Roman script for notes in ink, italic for notes in pencil, and a backslash for a line break. Some notes have also been added in square brackets by way of interpretation.

Folio 1, recto, (165x185mm)

Mr Boyd's; May, – 63?/ at Danson

Folio 1, verso (see Fig App1.1)

100 Yards, to an Inch/ (Mr Boyd's)/ Wly [Westerly]/ W^r Supply for so much/ Water?/ The copier(?) Road w^r practicable/ 300f. Far St: 600f KGn/ N [North]/ Not so sloping/ at 1200 f/ Thro ~~Lodge~~ *Wood/ to E Walks: (?????)/ & good.*

Folio 2, recto (325x205mm)

N: Front./ The Lawn-Hill$^{(2)}$, sloping/ down very pleasingly, & Longish ~~Slip~~ *left hand sd be first/* ~~of~~ $^{(1)}$ Wood, to the East, - Gentle/ Hill, more distant, & intersected/ by L: Hill, in the middle:/ – L: Hill, sinking, & several/ Woody Hills, (????) particular/ Shooter's-Hill with y^e London *Rises too near y^e H/ Road, to W.*

E: Front./ The intersected Hill, (as be-/ -fore,) to N; - Longish Wood, (ditto,) – *L Hill sloping down/* High Trees *(plot ye Ice House)* an^d ye Lawn, & Open-/ -ing to a Grand & Beautiful *Gentle land, & rising to House/* Prospect, in the Middle: - & / Home Wood, to S. (??) all desig^n, & risg (???) Lond(??)

Folio 2, verso

S: Front./ Longish Wood, High Trees,/ Part of the Fine Prospect/ (as before,) & other High Trees/ about the Old House, to E; -/ Water, & Home Wood, & distant/ Hills appearing above it, (par-/ -ticularly *Knockolt* Beeches.?) in/ the middle: - Wood, & more/ distant woody Hill, (with the/ London Road again,) to W.

W: Front./ Water, Home Wood, & pretty *above a/* Hill in the distant View to S. –/ - Wood, & Shooter's Hill (in-/-tersected ~~afterwards~~ $^{the\ Road}$ by Lawn/ Hill, which s^d be full planted,/ In the middle: - & the Gentle/ Intersected Hill, to N.

Folio 3 recto (325x205mm)

N: View./ The Bounds there to be/ more varied, & conceal'd: and/ knole of the Hill full planted/ to conceal it's shortness or/ lower'd to let in the Prospect.

E: View./ A Grove just w^h the Boundary/ to join the Hanger Wood, to/ the Eye, & to take off the strait-/ -ness of the line. – Trees to be/ taken away, where very preju-/ -dicial to the Grand Prospect.

S: View./ The greatest bredth of/ Water to be opposite to the S./ Front of the House; & to/ narrow and widen again –/ towards it's end. – To follow/ the Natural Fall of the/ ground in conducting the/ water, as much as can now/ be done. – To open the back/ wood, largely, & variously.

W: View./ Whether not too much blockt/ up by the Plant^n in Mr Rich-/ -mond's Plan from Wellend Grove/ down to the little meadows.

Folio 4, recto (285x380mm) (*see* Fig 2.7)

The Great Lawn; &c: about Mr Boyd's/ New House, at Danson

N^o		A.	R.	P.
1.	Clay's Wood	3.	1.	36.
2.	Cowspring Wood	3.	2.	28.
3.	Homespring Wood	26.	0.	16.
4.	Clay's Field	2.	2.	14.
5.	High Riddens	6.	1.	4.
6.	Great Chapel F^d	6.	2.	24.
7.	Chapel-Yard	0.	0.	16.
8.	Great Groves	5.	2.	16.
9.	Little Grove	1.	3.	26.
10.	South Lawn	19.	2.	22.
11.	Great Pond	2.	3.	1.
12.	Canal	1.	1.	12.
13.	North Lawn	97.	1.	10.
14.	Wellend F^d	10.	1.	29.
15.	Mudhouse F^d	7.	3.	32.
16.	Wellend Grove	4.	2.	31.
17.	Mudhouse, & G^n	0.	0.	35.
21.	Little Chapel F^d	1.	3.	13.
23.	Danson Mead:	1.	2.	3.
24.	Cherry Orch^d	6.	1.	33.
25.	H & G^n on Danson Gr:	0.	2.	8.
26.	Hangers Field.	2.	3.	5.
27.	Hangers Wood.	12.	2.	17.
30.	Gosweller's F^d	3.	0.	28.
31.	Ditto's	4.	2.	15.
33.	Litchbyrd's Mead:	0.	1.	37.
34.	Gough's Meadows	0.	2.	30.

APPENDIX 2

Sale Inventory of 1805

A transcript of the handwritten inventory, apparently by John Boyd II, dated 1805 and now in the collection of the late William Charter, Lee-on-Solent.[1] These items were left over after two auctions of Sir John Boyd's picture collection (1800 and 1805) and do not include items that the younger Boyd had decided to keep.

The Hall

The 2 jasper tables in the Hall	40
A marble statue of Adonis with a stone base and deal painted pedestal	75
1 ditto of Venus [both 'brought from Rome']	
2 ditto of plaister	5
A medusa head in statuary marble	8
A marble bust of Lord Chatham	35
A ditto of Lord Camden	
2 therms to support them	

In the [Dining] Parlor

The 3 bronze figures supporting an oriental spar cistern on marble splinth [sic]	57
A fine French urn for ditto	
2 statuary marble figures in splinths [sic]	30
2 therms for ditto	
2 large glasses.	50

Library

4 pier glasses	60
2 statuary marble figures on splinths [sic]	30
A Venus crouching	15
5 bookcases round the Room	12
The organ at your own price if the rest are all taken	105
2 mahogany chests of 5 drawers with mahogany bookcases on ditto	20
Silk curtains	

Temple

A carved marble figure of Venus bathing in a dove marble bath supported [sic].	25

Green House

A capital antique Vase supported by 3 figures on a marble pedestal dug from Adrians [sic] villa.	105

Drawing Room

Paintings: Verney [sic], Wilson, Barrett	250
A cut glass chandelier for 8 lights	14
A masque of marble	5

The old stone that is on the premises [from the previous house and demolished stables and offices?] the quantity is unknown and it is therefore proposed to leave it to a mason or builder who can say how many tons it contains – and what is the value per ton.

This may be had if the other things are rejected.

NB. There are some Bronzes not mentioned in this List.

The vases in the Green House and Temple to be taken at all events

NOTES

Abbreviations

BCRT   Building Conservation and Research Team
Bexley LSAC   Bexley Local Studies and Archive Centre, London Borough of Bexley
EH   English Heritage
GLC   Greater London Council
HART   Historical Analysis and Research Team

Preface

1. Tavernor and Schofield 1997, Bk 2, chap 12, 122; Ackerman 1990, 35–43.
2. Hancock 1995, 320–47, includes a discussion of the house-building habits of several City merchants of this period, including Boyd.
3. See Hancock 1995, 437–43 for 'A Catalogue of a Genuine Collection of Valuable Pictures by the Most Admired Masters of the Italian, French, Flemish and Dutch Schools, … the Property of Sir John Boyd, Bart., Deceased, By whom they were collected during a Long Residence in Italy, and now removed from his Seat in Kent' for sale at H Phillips, London, 18–19 Mar 1800.
4. Hancock 1995, 117.

1 The origins of Danson

1. Hutcherson 1996, 1–6.
2. Strype 1821, 467–8.
3. See Boyd's Private Act of Parliament, 2 George III, no 85 (1762).
4. For Kent in this period, the best source is Hasted 1797–1801. Danson is discussed in vol **2**, 172–3.
5. Greenhalgh 1995 (unpub), 5–7, 12, 33–7.
6. Wooller 2000.
7. Miele 1999, 31–5. Detailed survey information on suburban estates can be found in VCH Middlesex, **5**, **7** and **10** and VCH Essex, **6**. For South Kensington see Survey of London 1986.
8. Bexley LSAC Danson Leases, PEDAN/1/22 (Assignment 4 Jul 1695). See also Piper 1974 (unpub), 7–10.
9. Bexley LSAC 333.541 MAN (Survey of Manor of Bexley 1681) and copy of 1684 Map.
10. Bexley LSAC PEDAN/2/1 (Map 1710).
11. Jacques and van der Horst 1988, 142–3; Thurley 2003, 205–6.
12. John Styleman was the son of Robert Styleman, a merchant attached to the East India Company. He was made a secretary at the Company's Bantam factory in Java, 1669, and in 1679 became factor at Tonquin, North Vietnam. From 1685–8, he was trading as a free merchant under license from the Company in Madras, India, which was formally settled by the British in 1688. Allegations of financial double-dealings did not stop Styleman from serving as Mayor of Madras from 1692 to 1695. For more on John Styleman's career, see Wooller 2000, 56–60.
13. Piper 1974 (unpub), 10–25.
14. Ibid, 14–16; Bexley LSAC Danson Leases PEDAN/1/53 (Lease 20 Sept 1723).
15. The information here is taken from Sedgwick 1970, **2**, 415–16 and Carter 2004. Matson House is described in Oswald 1950, Bazeley 1877–8 and Jennings 1983, 8–9.
16. Kingsley 1989, 131. Selwyn indicates he is writing from 'Danson' in British Library MSS Add 35586 fol 196 (20 Sept 1739) and Add 35603 fol 55 (3 Nov 1748).
17. Piper 1974 (unpub), 15–16, 22; Bexley LSAC PEDAN/1/66–7, 71–4, 78–9 (Leases 1739–46).
18. This can be inferred by comparing the lease plan of 1753 (see Fig 1.4) with one of 1745 (Bexley LSAC PEDAN/2/2 (Map 1745)).
19. Bexley LSAC PEDAN/1/90 (Lease 10 Jul 1753). At that date the address of the freeholder, Selwyn's widow, Mary, is given as St George's, Hanover Square. John Boyd had already taken up residence by this point.
20. White 1982. Mr White located this drawing in Oxford, Bodleian Library, Gough Maps 13, fol 28.
21. Wright was patronised by the Townshend family, with whom Selwyn was allied. He taught mathematics to Lord Townshend's daughter, Elizabeth, and had worked for Townshend himself at Raynham Hall, Norfolk, and at Frognal, Kent, for Townshend's younger son, Thomas. Frognal is only two miles south of Danson, and is immediately adjacent to Scadbury, the property of Selwyn's mother. Also, Thomas Townshend was married to Selwyn's daughter Albinia; see Burke's Peerage … 1959, 2242. The domed temple on Spence's plan of Danson Park is similar to one designed by Wright in 1742 for the garden at Culford, Suffolk, in 1742, now in a volume of about 175 drawings at the Avery Library, Columbia Uni, New York, titled, 'Various & Valuable Sketches and Designs of Buildings' (call number AA543 W93 W933). We are most grateful to Dr Eileen Harris for this and other information about Wright's work.
22. Hughes 1951, 28.

2 Augustus and John Boyd

1. Hancock 1995, 33, 40–42, 47–53, 106, 115, 131, 143–7, 172–83, 189–207, 213–15, 218–23, 241–5, 253–4, 271–4, 285, 287–92; Leland and Ressinger 2006 and 2008. Jean Boyd's letters to his sister are among the papers of James Fraser (1645–1731), the Royal Librarian to Charles II and William III from 1685–92, now in the university library collection at Aberdeen, MS K 257/44/9.
2. Sheppard 1998, 129–47.
3. Hancock 1995, 172–220.
4. Rate books for the Parish of Lewisham, Lewisham local studies collection, list both Boyds in Ladywell, adjoining Lewisham High Street, for this period.
5. Philips 1941.
6. The development of Great George Street is discussed in Survey of London 1926, 12–14 and 21 (for No 4) and plate 17. Boyd was granted a lease on 21 Jun 1760,

where he was described as being 'of Lewisham' not the City. Number 4 was demolished in 1910 to make way for the Institution of Civil Engineers, and some original features were removed to the Victoria and Albert Museum. *See also* Bradley and Pevsner 2003, 273–4.
7   Garnier 2002b; Colvin 2008, 1028.
8   Binney 1984, 49; Colvin 2008, 1024.
9   Information on the Danson leases in this and the next two paragraphs comes from Piper 1974 (unpub).
10 Bexley LSAC PA23/8/A/1 (Bexley Vestry Minutes 1762–90).
11 Private Acts, 2 George III 1762, no 35.
12 Bexley LSAC PEDAN/2/4 (Map 1763).
13 Fisher 1776, 19–20: 'The disposition of the former is striking and beautiful; and when he examined the latter he will not fail to pay a compliment to Mr Brown's superior skill in forming and securing so large a piece of water.' Hasted 1798, 172–3: 'Behind the house, at a proper distance, is a most magnificent sheet of water, so contrived as to seem a beautiful serpentine river, flowing through the grounds. It was designed, and with much difficulty formed and secured by the noted Capability Brown, who likewise laid out the adjoining grounds, which are well cloathed with many thriving plantations of different kinds of trees.' Brayley 1808, vol 7, part 1, 556–7: 'The grounds were laid out by the celebrated Brown'.
14 The drawing (*see* Fig 2.5) is titled: 'A PLAN/ for the Alterations proposed/ at DANSON in KENT/ The Seat of/ John Boyd Esqr.' It measures about 880 x 1140mm and is drawn at a scale of 100ft to 1in, or 1:1200. It is neither signed nor dated. Features on the plan are labelled with numbers which are explained in the key titled 'References':
    1. The Approach to the House
    2. The House
    3. The Road to the Stables &c &c
    4. Situation for the Stables
    5. The Melon Ground
    6. The Kitchen Gardens
    7. The South Border
    8. The Ice-Well
    9. The Sunk Fence intended
    10. A Gravel Path
    11. A small Stew
    12. The intended Water where the House stands
    13. The Water to be altered
    14. A Riverlike Continuation of D^o.
    15. D^o.
    16. Langton's Wood
    17. Wood belonging to M^r. Boyd
    18. Langton's Field
    The plan was first noted in Jacques 1983, 84–5. *See also* Brown 1998–9, 37 and Brown 2001, 7–8.
15 Morris 1734, 161.
16 Worsley 2004, 132.
17 Ware 1756, 406.
18 Comparable wings are found on James Paine's design for a house at Gosforth, Northumberland, as published in Paine 1767, **1**, plate 20.
19 Joseph Spence's original notes on Danson consisting of a survey plan of the estate, a sketch plan for the water supply for the proposed lake and two sets of notes, are in the Beinecke Rare Book Library, Yale University, New Haven, USA: Osborn Shelves, Spence Papers, MSS 4/6/171. The plans and the notes are reproduced here in Appendix 1. Some of the text has already been published in King 1980b. Further notes on Spence were published in King 1978, 1979 and 1980a.
20 Wright 1950, 114–119, 136, 147.
21 'fairly' because the date on Spence's notes, though in his own hand, is a pencil addition which includes a question mark, 'May 63?'
22 *The Ambulator* ... 1782, 69.
23 C E Thorpe's drawing, dated Sept 1768, shows this folly newly built (British Library MSS Add 32353, fol 247).
24 Examples of Taylor's Gothic style can be seen in the spire of St Peter's Church, Wallingford, Berks (1776–7, demolished 1880); Long Ditton Church, Surrey (1778–9); and alterations at the Bishop's Palace, Salisbury, Wilts (1783–5).
25 Lea 2003b (unpub); Potter 2003 (unpub).
26 Tavernor and Schofield 1997, Bk 2, chap 12, 121–2, where Palladio also advised putting a house beside a river or on elevated ground at a distance from standing waters.

3   Robert Taylor, architect of Danson

1   For a brief outline of Taylor's life *see* Colvin 2008, 1023–8.
2   Binney 1984, 39–54 and 105–6 for bibliographical sources. Binney was also the first writer to examine Taylor's villas as a group; *see* Binney 1967.
3   Sir John Soane's Museum, Lincoln's Inn Fields, London, has a complete set of *The Architectural Designs of Sir Robert Taylor drawn and executed in aquatint by Thomas Malton* (1792). Several of the original drawings are held in the Ashmolean Museum.
4   The attribution was confirmed by Hasted: 'The original design for this structure was given by the late ingenious Mr. Taylor, architect of the Bank' in Hasted 1798, 172–3.
5   The Taylorian Institute has three volumes of Taylor drawings (chimney pieces, geometrical diagrams and monuments). For the books from Taylor's library at the Taylorian Institute *see* Gilson 1973.
6   The chief biographical source on Taylor's life is Horace Walpole's obituary published in *The Gentleman's Magazine* for 1788 (Walpole 1788a). A more detailed version was published in the *London Magazine* (Walpole 1788b). *See also* Binney 1984, 24–5.
7   Summerson 1993, 343–4.
8   Binney 1984, 13–15, 20–1.
9   Ibid, 28.
10 Summerson 1993, 344.
11 Andor Gomme attributed Barlaston to Taylor in Gomme 1968.
12 Binney 1984, 13.
13 Colvin 1954, 601–4.
14 Binney 1984, 94–7.
15 Colvin 1995, 962–7.
16 Colvin 2008, 1023–8.
17 For Richard Garnier's publications on Robert Taylor in the *Georgian Group Journal* between 1997 and 2008 *see* bibliography.
18 Girouard 1978, 183–192; Stillman 1988, 141–2.
19 Survey of London 1960, 192–202, 287–302.
20 Stillman 1988, 144–5; Worsley 1991.
21 Garnier 1997, note 1.
22 Binney 1984, 17–19.

4   Proportion and structure in Taylor's villa at Danson

1   Morris 1750 plates 2, 5, 30, 37. Taylor owned a copy of Morris 1750, *see* Gilson 1973.
2   For Watlington Park, Oxon (c 1760), Asgill House, Richmond (1761–4) and Ottershaw Park, Chertsey (soon after 1761), *see* Binney 1984, 46–7. In the 1750s, Stiff Leadbetter had used canted bays prominently at Langley Park, Bucks and Nuneham Park, Oxon (1755–6), *see* Worsley 1995, 236–40, figs 254 and 262.
3   Ware 1756, **4**, chap 16, 460, also recommended double-square windows: 'the height of windows for the principal storey is to be proportioned to their breadth ... the more general proportion in plain windows twice the measure of the aperture in breadth for its height'. Palladio states that windows should be two and a sixth times higher than their widths, leading to a more attenuated form, *see* Tavernor and Schofield 1997, Bk 1, chap 25, 60–1. Hoppus 1737, plate 46, identified the 1:2 window proportion as merely one of six equally acceptable options.
4   Tavernor and Schofield 1997, Bk 1, chap 16, 32–9.
5   Vitruvius 1914 edn **6**, chap 3, paras 7 and 8, 179.

6 The stone used for the strings and offsets is from Taynton, near Burford, Oxon, *see* Sanderson 1994 (unpub).
7 For Barlaston *see* Binney 1984. For Harleyford see Cornforth 2001.
8 Groves 2002 (unpub).

5 Completing the house and landscaping the park, 1756–1773

1 Thomas Malton taught perspective and published on the subject, *see* Malton 1778. That he was working from original design drawings, rather than from the house itself, is suggested by his rendering of the perron, which is shown narrower than built.
2 Watkin 1996, 620.
3 Lea 2003b (unpub).
4 Bexley LSAC MS 1/1–2 (Window tax assessments for 1766–7).
5 Hasted 1798, 162–83.
6 Fisher 1776, 19–20.
7 The console bracket carved with acanthus leaves is similar to those from beneath the vaults in one of the four Transfer Offices at the Bank of England (1765–8), now in Sir John Soane's Museum. Taylor continued to work as a sculptor after taking up architecture, and his last known commission is the monument to Edmund Auberry at Pinner, Middlesex, in 1767, *see* Binney 1984, 24, 72–3. In the completed interiors at Danson the acanthus console recurs in the woodwork carving of the organ case (*see* Fig 6.55).
8 Harris and Snodin 1996, 107–24.
9 Elias Martin was probably the courier who brought Vernet's painting for the saloon from Paris (*see* chap 6 p52). Vernet would have known that Martin and Chambers had a Swedish background in common; *see* Harris 1970, 12, 14, 20 and Arts Council of GB 1963, 11.
10 John Boyd to William Chambers, 11 Jun 1770, Royal Academy of Arts, London, CHA/1/19. See also Harris 1970, 203.
11 William Chambers to John Boyd, 12 Jun 1770, British Library MSS Add 41134, 24 and 25.
12 Harris 1970, 203.
13 GLC Historic Buildings File, AR/HB/704 (I), now EH Registered File, cited hereafter as EH AR/HB 704 (I), items dated 21 and 27 Jan 1960, 18 Jan, 10 and 15 Feb 1961 and 31 Dec 1963 (confirming that the Temple was removed in 1961).

6 Planning, decoration and iconography

1 Harleyford (1755), for example, has a double-height kitchen, as does 4 Grafton Street (1768–75).
2 The window in the northern part of the east wall was originally blind, like its counterpart to the south. It was opened up, and a groin cut through the vault, when the window in the north wall was blocked by the additions of *c* 1805.
3 This was partially observed in excavation, *see* Lea 2003b (unpub).
4 Skaife 1774, 183.
5 In the original scheme, the ground-floor hall was probably not directly linked to the double-height kitchen. The present stair in the kitchen dates from the 19th century and its insertion has removed all traces of the original treatment of this part of the fabric.
6 These rooms had simple plaster cornices, consisting of a cavetto moulding (concave quarter circle) above a small cyma-reversa (s-profile moulding, convex above, concave below). The chimney piece in the northern of the two bedrooms has an egg-and-dart timber surround with marble slips, and is similar to those on the bedroom floor in rooms of the same size. That in the southern room, on the other hand, while being grander in design, has no enrichment and is much larger despite being for the same room size. These characteristics alone suggest that it may not be in its original position.
7 Binney 1984, fig 63.
8 Originally there was a latrine in the inner angle. The door between the closet and the saloon was inserted after 1860, when this space became a linking corridor between the hall and the saloon. It was cut through the original brick of the north-east wall of the saloon, and the plaster made good around a recess framed to receive new mirrors, *see* Lea 2002c (unpub).
9 Binney 1984, figs 30, 36, 69, 70, 75.
10 These may have been 19th-century additions, but more probably were original and simply not mentioned in 1805 because they were deemed to be fixtures. A copy of the 1922 sale catalogue, *The Danson Estate, Kent, 1922*, is held in Bexley LSAC (728.8DAN).
11 Hughes 2002 (unpub).
12 Traces of various hinges, observed before the mahogany door was restored, indicate that the door has been altered and rehung many times, but the presence of the original shutters demonstrates that the upper part of the door was originally glazed. That the glazed area was not subdivided by glazing bars is evident from the egg-and-dart surround which shows no sign of mortices for glazing bars. Part-glazed doors can be found at several of Taylor's villas but the glass is always subdivided.
13 The present lantern was installed in 2004.
14 Thieme and Becke 1922, **26**, 321.
15 Hughes 2007 (unpub). The floral panel on the south side of the bay is exceptional in being glued to the wall with a mixture of lead white and glue size. Edward Croft-Murray (1970, 324–5) briefly describes the cycle (although his iconographic analysis is incorrect).
16 Hancock 1995, 350, note 26. Andrea Casali, who provided canvases for the octagonal room in Taylor's Asgill House in around 1762, socialised with house guests when decorating Copt Hall in the 1740s. In 1743 Lady Pomfret commented: 'After supper we danced to our own singing in order to teach Signor Casali (an Italian they have in the house) English country dances … . He is a painter and I fancy as low born as they generally are, though by means of an order he wears (which he tells them was given by the King of Prussia, and which very few people have) and some fine suits of clothes, he passes for the most complete country gentleman in the world. They tell me he paints more in two hours than any other of his profession can do in a day', *see* Hussey 1944, 994.
17 We are grateful to John Hardy of Christie's, London, for directing us to Virgil's Georgics as a source; *see* Day Lewis 1974, 51–87.
18 Harris 2001, 161.
19 Stuart 1762 **1**, chap 4, plate 3.
20 Hussey 1956, 113–14.
21 Ovid, *Metamorphoses*, 14, 623–97, 765–71, trans Miller 1916.
22 The senses were allegorically depicted in a painted scheme of Charles Simon's, representing Apollo and Cybele, in the dining room of Canons House, Middlesex, completed around 1725, *see* Croft-Murray 1962, no 276b.
23 Dr Celia Fisher, author of *Flowers in Medieval Manuscripts* (London: British Library 2004) and *The Medieval Flower Book* (London: British Library 2007), advised EH on the plant and flower species in the Pavillon paintings in a report in 1997.
24 Inventory in the collection of the late Commander William Charter, Lee-on-Solent; at the time of going to press (2011) the future location of this archive had not been determined.
25 One of the Sarah Johnston watercolours shows the room furnished with a sofa (*see* Fig 6.14).
26 Tavernor and Schofield 1997, Bk 1, 46–7. The depth of Taylor's modillion course is slightly reduced and the dentil and ovolo mouldings correspondingly enlarged. The arabesque decoration was removed in 1937. It was restored by EH from photographic evidence.

27 Printed catalogue for auction of 27 June 1805, 5, in the collection of the late Commander William Charter (see note 24, above). There is a copy of this catalogue in EH file HA 054094/01 PC46.
28 The roll width of the new paper, 22in, is exactly one sixth of the 11-ft width of each wall of the octagon. This greatly facilitated the hanging of the paper, especially the positioning of each roll in relation to the corners. In 2003, 22 ins was the largest width that Adelphi Paper Hangings could make. The spacing of the anthemion in the frieze also fits exactly with the 22-in width. This suggests that the size of the octagonal room and the design of the anthemion frieze were dictated by the desire to use wallpaper with a 22-in roll width. Octagonal saloons at Taylor's other villas – Asgill, Sharpham, Chute, and Mount Clare – are all the same dimensions, each having walls 11-ft wide, suggesting that all these rooms were designed to be lined with paper or fabric in 22-in widths. Although the glue residue showed us that in the Danson saloon the wall was originally hung with a paper or fabric with a 26-in width, this dimension is almost exactly one fifth of 11ft, so again the hanging of the wallcovering would have been made easy.
29 Cooper-Hewitt National Design Museum, New York, 1921–6–D198 and plate 26 in Pillement 1773. In 2000, the textile historian Mary Schoeser advised EH that the original wallcovering would have been wallpaper, and that the damask described in 1805 was probably either flock paper imitating damask or a replacement wallcovering from c 1805. Like the original, the new paper is hung on a hessian scrim stretched across the wall face. It is because some of the glue used to hang the original paper over time migrated to the plaster wall face that we have a partial image of the original design. Traces of glue from the original paper showed that the width of its pattern was 26in. To match the scale of the pattern, the Cooper Hewitt paper pattern was enlarged by more than 100 per cent. See also Schoeser 2009.
30 Binney 1984, 46.
31 Lagrange 1864, 247: '[Order number] 229.Tableau pour M. Boyd anglois à Londres, ordonné par M. Vanloo par une lettre qu'il a reçû de M. Pavillon. Il doit avoir cinqs pieds de large, sur six pieds et un pouce de haut mesure d'Angleterre prise sur le pied anglois. Il doit representer une grande chutte d'eau, des lointains et orné de beaucoup de figures; le prix est de cent cinquante loüis ou 3600 l. Je l'ay promis pour le mois de mars de l'année 1768. Il a été ordonné en décembre 1766'; and 'Par une lettre de M. Pavillon du 24ᵉ février 1767 ecritte a M. Vanloo, il a envoyé une nouvelle mesure du tableau cy dessus qui est 5 pieds 10 pouces de haut, sur 4 pieds 6 pouces de large mesure d'Angleterre, ou 5 pieds 5 pouces 10 lignes de haut, sur 4 pieds 2 pouces 3 lignes de large mesure de France, il doit y avoir toujours des chuttes d'eau mis avec un fond de marine. Les deux dessus de porte du salon ou doit être ce tableau ettant en paysage. Tout le reste a l'égard dudit tableau comme cy dessus. [Receipt number] 137'.
32 See Lagrange 1864, 365: '[Reçûs] 137. Vers la fin d'avril 1768 j'ay reçû de M. Vanloo 2400 l. qui avec 1200 l. que j'avois deja reçû fait la somme de 3600 l. pour le prix d'un tableau representant un paysage et marine avec de grandes chuttes d'eau pr M. Boyd de Londres'.
33 Croft-Murray 1970, 243; Arts Council of Great Britain, 1963, 8.
34 Martin's painting was lot 46; see Hancock 1995, App VII, 437–43 and Lea 2003a (unpub), 29.
35 Patron and painter were involved in other dealings at this time. A fine, delicate watercolour in the Guildhall Library by Martin shows Blackfriars Bridge under construction (around 1767), and is inscribed 'Painted for [Sir] John Boyd and the architect of the bridge', that is, the engineer Robert Mylne. Martin was clearly taken into Boyd's circle of acquaintances and he must have met his architect. Taylor, besides his connection at Danson, was consultant to the bridge's construction and also advised on the legislation that empowered it.
36 John Boyd II's inventory of 1805 states that one of the three paintings in the saloon was by Barrett, one by Vernet, and the third by 'Wilson' (see Appendix 2). However, in a letter of 9 Sept 1806 to John Johnston, the next owner of Danson, Boyd II refers to two paintings of Danson by Barrett, an inconsistency which calls into question his earlier mention of Wilson (see chap 7 p67). For a more detailed discussion of the attribution and an analysis of the specially commissioned saloon paintings see Lea 2003a (unpub).
37 1762mm by 1378mm (5ft 9⅜ins high by 4ft 6¼ins wide).
38 Nail holes in the timber battens surrounding the fireplace opening show that the saloon was lined with wallpaper or damask scrim before the Chambers chimney piece was fitted. The overall width to the edges of the wallpaper on each side was 5ft 1 inch, thus allowing 4 ins of frame on each side of the painting together with ½in overlap between the frame and the picture.
39 Sir John Soane's Museum, drawer 42/3/8.
40 V & A Museum Drawings Collection, nos 7078.2, E4984–1910, and 3861.19 and 20; see also Snodin 1996, 132–3, cat nos 623–6.
41 New York, Columbia University, Avery Architectural Library, IC/2 9; see also Harris 1970, 224.
42 de Montfaucon 1719.
43 The story of Cupid and Pysche is taken from Books 3 and 4 of Apuleius' *Golden Ass*. Psyche was a beautiful young woman who made the goddess Venus envious. As punishment, Venus sent the god Cupid to arouse the mortal's passion in someone completely worthless, but the Cupid was himself overawed by Psyche. They became lovers, and Cupid made Pysche promise never to look on him. One night, however, she raised a lantern above his slumbering body and woke Cupid. At once, he and the palace disappeared. Forlorn and inconsolable, Pysche wandered the earth searching for him, performing ever more difficult tasks set by Venus. Eventually, Jupiter intervened and raised Psyche to Olympus where she married Cupid in the presence of the pantheon of gods; de Montfaucon 1719, **1**, plate 121, no 1 and 192.
44 In line with the other Danson interiors, verdigris bespoke wealth. In about 1811, John Pincot noted (Pincot c 1811) that while common colours cost 3½d per yard, 'verdigris green … or any very extravagant colours (cost) 8d per yard extra.' See also Hughes 2002 (unpub).
45 Gilbert 1978, 98.
46 Ibid, 44.
47 Joy 1971, plates 90, 91 and 95. (Plates 88 to 101 of Joy are devoted to bookcases.) See also Claxton Stevens and Whittington 1983, 190, 200.
48 The bookcases at Danson were adapted to take glazed doors in the late 19th century. Sarah Johnston's watercolour of the bookcase against the north wall shows it before the conversion (see Fig 6.56).
49 Langley 1756, plate 158.
50 The central plaque was removed from this location and hung in the closet next to the stairs, probably in around 1805. It was restored to its present position in 1997. The ribbons were reconstructed on the basis of their outlines, which were preserved in paint on the plaster wall face above the chimney piece.
51 Bartoli 1693. The same image is also illustrated in Barbault 1761 as an untitled plate between pages 14 and 15, where it appears as a stone fragment. The timing of this last publication may relate to the

sudden appearance of plaster copies in this country in the 1760s.
52 The design itself is unlikely to have been produced before 1762–3, when General Burton returned a hero from the French and Indian War in Canada, and must predate his death in 1768, see Hall 1978–9, 77–8, fig 100. In Harris 1970, 246–7, the drawing is associated with Slane Castle (Co Meath, Ireland), where Chambers produced designs for Henry (Burton), Viscount Conyngham, in the 1760s; but Harris acknowledges that it is 'not entirely clear when Henry Conyngham could have been titled General …'.
53 See Binney 1984, fig 11, for Malton's view of one of the four Transfer Offices at the Bank of England (1765–8), where similar acanthus brackets support Corinthian wall capitals. The organ was removed to Hall Place in the early 1970s, where it was used for concerts in the great hall. It was restored and returned to Danson in 2003.
54 George England built organs for Dulwich College Chapel (1760) and St Stephen's Walbrook in the City (1765). When he retired in 1766 he was succeeded by his brother John whose earliest surviving organ is at Wardour Castle, Wiltshire (1767). In about 1790 John England was succeeded by his son, George Pike England, who continued the business until his death in 1815. It seems unlikely that either George or John England would sign their own work with the moniker 'Old England'. The organ restorer Noel Mander has suggested that the label was added when the organ was overhauled in the early 19th century, possibly by George Pike England. He also argues that George Pike may have chosen to refer to his father John as 'Old England' to distinguish him from himself. The name is equally applicable to George England, George Pike's uncle, who was probably the older of the two brothers. For Mander's correspondence, see Bexley LSAC LABX/DA/4/1/97. On English organs and the Danson instrument specifically, see Bicknell 1996, 179, 203 and 215; and Wilson 1968, 64–5.
55 Wilson 1968, 67.
56 See Harris c 1987, 36–7; Beard 1978, plate 58; Harris 1995.
57 Lea 2002b (unpub).

7 The cost of life as a gentleman

1 Hancock 1995, 33 note 25; Cockayne 1900–6, **5**, 184.
2 Hancock 1995, 350, note 28.
3 Piranesi 1778, **2**, plates 58–9: 'Vedute di Prospettiva per angolo di un'antico Vaso di Marmo con suo Piedestallo, che al presente si vede in Inghilterra nella Villa del Sig. Giovanni Boyd. Fu ritrovato negle Scavi fatti nella Villa Adriana l'anno 1769'. Hamilton 1901 refers to the vase in a letter to Charles Townley. Hancock 1995, 350, note 28 identifies Charles Townley as the restorer. Known as the Piranesi Vase, it is now in the King's Library at the British Museum. Piranesi himself may have been responsible for its restoration, see Jones 1990, 133.
4 The two principal beams in the floor run north–south. It might be argued that the partition walls subdividing the ground-floor entrance area into three rooms strengthen the floor above but they would not have reinforced the floor joists in the centre of the room.
5 We know this from a series of letters written by John Boyd II to Sir Robert Murray Keith between 1775 and 1782, cited below, notes 6–10 and 20. Keith was the British ambassador at Vienna and had taken on the younger John Boyd as an assistant, a post that, being unpaid, was only open to someone of independent means.
6 British Library MSS Add 35517, fol 3 (4 Jul 1779).
7 John Boyd II to Keith, 9 Dec 1780, in British Library MSS Add 35520 fol 138.
8 John Boyd II to Keith, 8 Feb 1780, in British Library MSS Add 35518 fol 63.
9 See correspondence from John Boyd II to Keith in 1779 and 1780, in British Library MSS Add 35516 fol 68; Add 35517 fols 3, 89, 127, 162; Add 35518 fols 151, 178; Add 35519 fols 1, 167, 230; Add 35520 fol 138.
10 John Boyd II to Keith, 9 Dec 1780, in British Library MSS Add 35520, fol 138.
11 Piper 1974 (unpub), 49–60 and corresponding leases in Bexley LSAC PEDAN series.
12 Hancock 1995, 215.
13 Piper 1974 (unpub) 55–60.
14 By way of compensation, the main bedroom-floor beams were suspended using wrought-iron ties attached to the top plates of the partition walls, which in turn were let into the masonry walls spanning the bays at main cornice level.
15 Survey of London 1980, 212; see also Survey of London 1977, 121, fig 8b.
16 Binney 1981; Garnier 2003b.
17 GLC Historians' Report, Historic Buildings Division, WM261, letter of 1 Apr 1981 (present location unknown).
18 Garnier 2003b.
19 John Boyd II's bank accounts survive in the Drummonds Archive, Whitehall Branch, London (DR/427/172, 176 and 182, for the years 1802, 1803 and 1804 respectively). Disbursements are consistent with building works known to have been carried out on the estate at this period.
20 John Boyd II to Keith, 13 and 24 Apr 1779, BL MSS Add 35516 fols 95 and 123; Add 35517 fol 127 (30 Aug 1779); and Add 35519 fol 1 (1 Jun 1780).
21 Bexley LSAC, PEDAN/1/235–6 (Lease and release 22–23 Jul 1806), PEDAN/1/242–4 (Lease, quitclaim and release 3 Jul 1807) and PEDAN/1/250 217 (Assignment 14 Jul 1807).
22 Worsley 2004, 256–9.
23 There is the parallel example of Witley Court, Worcs, which had two free-standing service wings linked by quadrant walls to a principal front, the whole probably built in the late 1720s. The wings were demolished in 1804 and a new range combining stables and offices constructed to the west of the main house, just out of sight. All this was probably under the influence, if not direction, of John Nash who added porticoes to the house at the same time. See White 2003, 20–22.
24 Sale catalogue with attached estate plan (see Fig 7.3) in the collection of the late Commander William Charter, see chap 6, note 27. Copy of this catalogue in EH file HA 054094/01 PC46.
25 Bexley LSAC PEDAN/1/223–4 (Lease and release 5–6 Aug 1800).
26 Bexley LSAC PEDAN/1/226–7 (Lease and release 20–21 Dec 1802) and PEDAN/1/236 (Release 23 Jul 1806), where the payment is recorded in a receipt pinned to the legal document.
27 Sir John Soane's Museum, Dance Cabinet, D2/9/8; Worsley 2004, 190–1.
28 See chap 7, note 19.
29 Sale catalogue with attached estate plan (see Fig 7.3) in the collection of the late Commander William Charter, see chap 6, note 27.
30 See chap 6, note 27.
31 Bexley LSAC PEDAN/1/235–6 (Lease and release 22–23 Jul 1806).
32 Bexley LSAC PEDAN/1/250 (Assignment 14 Jul 1807).
33 Address given on Bexley LSAC PEDAN/1/236 (Release 23 Jul 1806).
34 In the collection of the late Commander William Charter, see chap 6, note 24.

8 Nineteenth-century Danson

1 We are very grateful to Commander William Charter, of Lee-on-Solent, for sharing his researches on the Johnston family with us.
2 Hutcherson 1996, 30–1. However it should be noted that, contrary to Hutcherson, Johnston was not a 'retired captain' who had served in the 62nd Regiment of Foot.

NOTES

3 Bexley LSAC PEDAN/2/5 (Map 1830). For Johnston's will *see* National Archives, PROB 1751, 88, dated 21 Mar 1828. Codicil of 1 Nov 1828. Johnston died on 24 Dec 1828, and the will was proved in Feb 1829.

4 *See below*, p77–8. The brick foundations of the orangery survive below ground level, *see* Potter 2003 (unpub).

5 The 1898 Ordnance Survey shows a lodge in the north-west corner of the park, now the junction between Bellegrove Road and Danson Crescent, and another small lodge on the east side, a few metres north of the junction between the present Danson Road and Bean Road. However, these lodges are not present on an earlier OS map of 1895.

6 White 1982, 66.

7 National Archives, IR 29/17/27.

8 Information from a descendant of the Bean family.

9 Information to Chris Miele from Christopher Date, Asst Archivist, British Museum, 13 Nov 1995.

10 Bexley's development after 1880 is detailed in Carr 1982.

11 1871 and 1901 Census returns.

9 Danson House and Park since 1924: decline and restoration

1 Carr 1970 (unpub); Carr 1982.

2 Mandler 1997, 225–63.

3 The price was £16,000, less than half the price John Johnston had paid more than a century earlier.

4 Bexley LSAC P914BEX, 11.

5 Bexley LSAC 728.8DAN.

6 Bexley LSAC LABX/CA/1/2/6 (Bexley UDC Mins 1933–4) entry for 14 Nov 1933.

7 Bexley LSAC LABX/CA/1/2/6–7 (Bexley UDC Mins 1933–4) entries for 24 Apr and 16 May 1934. There was movement in the floor structure above the library, caused, primarily, by the removal of structural timber in the 1780s when the side bay was raised. Repair estimates and schedules were never implemented.

8 Bexley LSAC LABX/DA/4/1/61 (Room hire correspondence 1935–51), report of 18 Jan 1937.

9 Documents relating to these repairs have not been discovered. The works are noted in the Bexley LSAC LABX/CA/1/2/21–6 (Bexley BC Mins 1947–52) entries for 13 Oct 1947, 9 Feb 1948, 10 Jul, 18 Sept and 27 Nov 1950, 15 and 29 Apr 1952.

10 For a discussion of the emergence of the current statutory regime, *see* Delafons 1997, 55–115.

11 The organ had become inoperable during the War, and in 1957 Bexley UDC commissioned Noel Mander of the St Peter's Organ Works in Hackney to restore the instrument. He identified one very significant alteration: in 1855, during the tenure of Hugh Johnston's mother, a Sesqualtera Bass replaced the original Cornet Treble. Restoring this treble made the instrument more suitable for the performance of 18th-century organ music. *See* Bexley LSAC LABX/DA/4/1/97 (Organ correspondence 1951–64).

12 EH AR/HB 704 (I), items dated 21 and 27 Jan 1960, 18 Jan, 10 and 15 Feb 1961, 31 Dec 1963 (confirming that the Temple was removed in 1961). Survey drawings survive in this file. There is no record of when the timber bridge, built by Chambers, was demolished.

13 The documents relating to the stable block are found in EH AR/HB 704 (I), items dated, 15, 21, 30 Oct 1964, 17 and 19 Sept, 10 Oct, 13, 20 and 25 Nov 1969, [?] Mar 1970, 2 Apr 1970. Inspector's decision, 13 Aug 1970. The same file contains technical reports, correspondence, public inquiry proofs of evidence, and memoranda relating to the conservation of the stable block from 1970 to 1974.

14 EH AR/HB 704 (I), letters from GLC to London Borough of Bexley, 22 Jan 1970 and from London Borough of Bexley to GLC, 2 Apr 1970; and proofs of evidence of the Borough Engineer, P E Morris and the GLC's Historic Buildings Division Director, Philip Whitbourn.

15 EH AR/HB 704 (II), Jan 1986 and following.

16 EH AR/HB 704 (III), correspondence and memoranda, Dec 1990 to Feb 1991.

17 EH AR/HB 704 (III), correspondence and memoranda, Mar 1991 to May 1992.

18 EH AR/HB 704 (III), correspondence and memoranda, summer 1993.

19 The authors are very grateful to Jamie Coath for sharing his recollections of this project. This section is based largely on discussions with him.

20 The performance of the shelter coat and the desirability of its renewal has since been assessed, *see* Fidler and Stewart 2002 (unpub).

21 For example, in the entrance hall, the central Victorian light fitting was replaced by a new lantern based on a surviving example at Somerset House from the 1770s. In the dining room and library, the damaged late Victorian gasoliers were not reinstated because, in the Georgian period, there was no central light source in either room. Similarly, the fragmentary remains of papier-mâché ceiling rosettes, installed at the same time as the gasoliers, were also removed.

22 Lea 2002c (unpub), 8.

23 Lea 2002a (unpub).

24 Lea and Miele 2001 (unpub).

25 Publications on Danson around this time include Mead and Hughes 2004 and Miele 2005.

Appendix 2

1 In the collection of the late Commander William Charter, *see* chap 6, note 24.

BIBLIOGRAPHY

Published sources

Ackerman, J S 1990 *The Villa*. London: Thames and Hudson

Apuleius, L nd *The Golden Ass*, trans by W Adlington, 1996 edn. Ware: Wordsworth Editions Ltd

Arts Council of Great Britain, 1963 *Elias Martin, 1739–1818*. An exhibition organised by the Arts Council of Great Britain with the National Museum and the Swedish Institute for Cultural Relations, Stockholm. London: Arts Council of Great Britain

Barbault, J 1761 *Les Plus Beaux Monuments de Rome Ancienne*. Rome: Bouchard and Gravier

Bartoli, P S 1693 *Romanarum Admiranda Monumenta*. Rome: D de Rubeis

Bazeley, W 1877–8 'Some records of Matson in the county of Gloucester, and of the Selwyns'. *Trans Bristol and Gloucestershire Archaeol Soc* **2** 1877–8, 259–65

Beard, G 1978 *The Work of Robert Adam*. Edinburgh: J Bartholomew

Bicknell, S 1996 *The History of the English Organ*. Cambridge: Cambridge University Press

Binney, M 1967 'The villas of Sir Robert Taylor', pts I and II. *Country Life* **142**, (6 and 13 Jul 1967), 17–21, 78–82

Binney, M 1981 'Sir Robert Taylor's Grafton Street', pts I and II. *Country Life* **170**, (12 and 19 Nov 1981), 1634–7, 1766–9

Binney, M 1984 *Sir Robert Taylor*. London: Allen and Unwin

Bradley, S and Pevsner, N 2003 *The Buildings of England: London* **6**: *Westminster*. New Haven and London: Yale University Press

Brayley, E W 1808 *The Beauties of England and Wales* **7**: *Hertfordshire, Huntingdonshire and Kent*. London: Vernor and Hood

Brown, D 1998–9 'Nathaniel Richmond, one of the first ornamental gardeners and the London network in the mid-Georgian period'. *The London Gardener* **4**, (1998–9), 36–9

Brown, D 2001 'Lancelot Brown and his associates'. *J Garden Hist* **29** No 1, (Special issue, Summer 2001: 'Lancelot Brown (1716–83) and the landscape park'), 2–11

Burke's Peerage … 1959. London: Burke's Peerage

Carr, M C 1982 'The growth and characteristics of a metropolitan suburb: Bexley, N. W. Kent', in *The Rise of Suburbia*, Thompson, F M L (ed) Leicester: Leicester University Press, 211–59

Carter, P 2004 'Selwyn, George Augustus (1719–1791)' in *Oxford Dictionary of National Biography*. Oxford: Oxford University Press

Chambers, W 1759 *A Treatise on Civil Architecture …* . London: for the author

Chippendale, T 1754 *The Gentleman and Cabinet-Maker's Director …* . London: for the author

Claxton Stevens, C and Whittington, S 1983 *Eighteenth Century English Furniture: The Norman Adams Collection*. Woodbridge: The Antique Collectors' Club

Cockayne, G E 1900–6 *Complete Baronetage* [5 vols]. Exeter: W Pollard and Co

Colvin, H 1954 *A Biographical Dictionary of British Architects, 1600–1840*. London: John Murray

Colvin, H 1995 *A Biographical Dictionary of British Architects, 1600–1840*. New Haven and London: Yale University Press

Colvin, H 2008 A Biographical Dictionary of British Architects, 1600–1840. New Haven and London: Yale University Press

Cornforth, J 2001 'Books do furnish a living room'. *Country Life*, **195** (13 Dec 2001) 56–9

Croft-Murray, E 1970 *Decorative Painting in England, 1537–1837* **2**. London: Country Life

Day Lewis, C 1974 *The Eclogues, Georgics and Aeneid of Virgil*. Oxford: Oxford University Press

Delafons, J 1997 *Politics and Preservation: A Policy History of the Built Heritage, 1882–1996*. London: E and F N Spon

Fisher, T 1776 *The Kentish Traveller's Companion …* . Rochester: for the author

Garnier, R 1997 'Two crystalline villas of the 1760s'. *Georgian Group J* **7** (1997), 9–25

Garnier, R 1998a 'Gatton Town Hall'. *Georgian Group J* **8** (1998), 72–5

Garnier, R 1998b 'Arno's Grove, Southgate'. *Georgian Group J* **8** (1998), 122–34

Garnier, R 1999 'Downing Square in the 1770s and 1780s'. *Georgian Group J* **9** (1999) 139–57

Garnier, R 2002a 'The Office of the Sick and Hurt Board'. *Georgian Group J* **11**, (2002) 96–100

Garnier, R 2002b 'Speculative housing in 1750s London'. *Georgian Group J* **11**, (2002) 163–214

Garnier, R 2003a 'Broom House, Fulham'. *Georgian Group J* **13**, (2003) 168–80

Garnier, R 2003b 'Grafton Street, Mayfair'. *Georgian Group J* **13**, (2003) 201–72

Garnier, R 2004 'The Grange and May's Buildings, Croom's Hill, Greenwich'. *Georgian Group J* **14**, (2004) 261–86

Gilson, D J 1973 *Books from the Library of Sir Robert Taylor in the Library of the Taylor Institution, Oxford: A Checklist*. Oxford: Taylor Institution

Gilbert, C 1978 *The Life and Work of Thomas Chippendale*. London: Studio Vista: Christies

Girouard, M 1978 *Life in the English Country House: A Social and Architectural History*. London and New Haven: Yale University Press

Gomme, A 1968 'The architect of Barlaston Hall'. *Country Life*, **143**, (18 Apr 1968), 975–9

Hall, I and E 1978–9 *A New Picture of Georgian Hull*. York: Sessions for Hull Civic Soc

Hamilton, G 1901. Letters to Charles Townley published in the *J of Hellenic Studies*, **21**, 1901, 310, 313, 321

Hancock, D 1995 *Citizens of the World: London Merchants and the Integration of the British Atlantic Community, 1735–1785*. Cambridge: Cambridge University Press

Harris, E 1995 'The Williams-Wynn chamber organ by Robert Adam', in Phillips Sale Catalogue, (21 Apr 1995). London: Phillips

Harris, E 2001 *The Genius of Robert Adam: His Interiors*. New Haven and London: Yale University Press

Harris, J 1970 *Sir William Chambers: Knight of the Polar Star*. London: Zwemmer

Harris, J 2004 'Chambers, William (1723–1796)' in *Oxford Dictionary of National Biography*. Oxford: Oxford University Press

Harris, J and Snodin, M (eds) 1996 *Sir William Chambers, Architect to George III*. New Haven and London: Yale University Press

Harris, L c 1987 *Robert Adam and Kedleston*. London: The National Trust

Hasted, E 1797–1801 *The History and Topographical Survey of the County of Kent*. Canterbury Bristow (Vol **2** 1798)

Hoppus, E 1737 *The Gentleman's and Builder's Repository*. London: James Hodges and Benjamin Cole

Hughes E 1951 'The early journal of Thomas Wright of Durham'. *Annals of Science*, **7**, 28 Mar 1951

Hussey, C 1944 'Asgill House, Richmond Surrey'. *Country Life*, **95**, (9 Jun 1944), 992–5

Hussey, C 1956 *English Country Houses: Mid-Georgian, 1760–1800*. London: Country Life

Hutcherson, R 1996 *The History of Danson*. Bexley: Bexley Leisure Services Department

Jacques, D 1983 *Georgian Gardens: the Reign of Nature*. London: Batsford

Jacques, D and van der Horst, A J 1988 *The Gardens of William and Mary*. London: Christopher Helm

Jennings A 1983 *A History of Matson, Gloucester*. Gloucester

Jones, M (ed) 1990 *Fake? The Art of Deception* (exhibition catalogue). London: British Museum

Joy, E T 1971 *Chippendale (Country Life Collectors Guides)*. London: Hamlyn

King, R H 1978 'Joseph Spence of Byfleet [Part I]'. *J Garden Hist*, **6**, (no 3 winter 1978), 38–64

King, R H 1979 'Joseph Spence of Byfleet [Part II]'. *J Garden Hist*, **7**, (no 3 winter 1979), 29–48

King, R H 1980a 'Joseph Spence of Byfleet [Part III]'. *J Garden Hist*, **8**, (no 2 summer 1980), 44–65

King, R H 1980b 'Joseph Spence of Byfleet [Part IV]'. *J Garden Hist*, **8**, (no 3 winter 1980), 77–114

Kingsley, N, 1989 *The Country Houses of Gloucestershire, Vol 1, 1500–1660*. Cheltenham: N Kingsley

Lagrange, L 1864 *Les Vernet, Joseph Vernet et la peinture au XVIIIe siècle … avec le texte des Livres de Raison*. Paris: Didier

Langley, B 1756 *The City and Country Builder's and Workman's Treasury of Designs*. London: S Harding

Leland, H C and Ressinger, D W 2006 *'Ce Païs Tant Désiré' ('This much longed-for country')*. Trans Huguenot Soc of South Carolina, **110** & Supp to **110**, 1–43; 2008 further Supp to **112**, 76

Malton, T 1778 *A Complete Treatise on Perspective in Theory and Practice*. London: for the author

Mandler, P 1997 *The Fall and Rise of the Stately Home*. New Haven and London: Yale University Press

Mead, A and Hughes, H 2004 'Back from the dead, Danson House, England'. *Architects' J*, **220** (19) (18 Nov 2004) 26–37

Miele, C 1999 'From aristocratic ideal to middle-class idyll', in *London Suburbs*, edited by Julian Honer. London: Merell Holberton in assoc with EH, 31–59

Miele, C 2005 'Danson House, Bexleyheath, Kent'. *Country Life* **199**, (24 Mar 2005), 94–9

de Montfaucon, B 1719 *L'Antiquité expliquée et représentée en figures*. Paris (trans into English in 5 vols 1721–5 by D Humphreys). London: Tonson and Watts

Morris, R 1734 *Lectures on Architecture. Consisting of rules founded upon harmonick and arithmetical proportions in building*. London: J Brindley

Morris, R 1750 *Rural Architecture Consisting of Regular Designs of Plans and Elevations for Buildings in the Country*. London

Oswald, A 1950 'Matson House, Gloucestershire'. *Country Life*, **108**, (8 Dec 1950) 1990–4

Ovid, *Metamorphoses*, **14**, 623–97 and 765–71 (trans into English by F J Miller 1916. London)

Paine, J 1767 *Plans, Elevations, and Sections of Noblemen and Gentlemen's Houses …* . London: for the author

Palladio, A 1570 *I Quattro Libri dell'Architettura*. Venice (trans into English by Isaac Ware 1738. London: Isaac Ware)

Pearce, D 1991 'The Naval and Military Club'. *Country Life*, **185**, (20 Jun 1991), 100–3

Philips, C H and D 1941 'Alphabetical list of directors of the East India Company from 1758 to 1858'. *Journal of the Asiatic Society*, Oct 1941

Pillement, J B 1773 *Nouveau Cahier de six feuilles… . Sujets Chinois Histoires*. Paris

Pincot, J c 1811 *Pincot's Treatise on the Practical Part of Coach and House Painting …* . London

Piranesi G B 1778 *Vasi, Candelabra, Cippi, Sarcofagi, Tripodi, Lucerne, ed Ornamenti Antichi*. Rome: Piranesi

Schoeser, M 2009 'The Octagon Room at Danson: evidence for a restoration with wallpaper', in Stavenow-Hidemark, E *New Discoveries, New Research, Papers from the International Wallpaper Conference at the Nordiska Museet, Stockholm*. Stockholm: Nordiska Mussets Förlag, 70–87

Sedgwick, R 1970 *The House of Commons 1715–1754*. London: HMSO

Sheppard, F 1998 *London: A History*. Oxford: Oxford University Press

Skaife, T 1774 *A Key to Civil Architecture; or, The Universal British Builder, etc*. London: I Moore & Co

Snodin, M 1996 *Sir William Chambers* (exhibition catalogue V&A). London: V&A Publications

Stillman, D 1988 *English Neo-classical Architecture* **1**. London: A Zwemmer

Strype, J 1821 *The Life and Acts of Matthew Parker…* **2**. Oxford: Clarendon Press

Stuart, J and Revett, N 1762 *Antiquities of Athens*. London

Summerson, J 1993 *Architecture in Britain, 1530–1830*. New Haven and London: Yale University Press

Survey of London 1926 **10** *The Parish of St Margaret's, Westminster* Part I. London: Batsford

Survey of London 1960 **29** *The Parish of St James Westminster* Part I. London: The Athlone Press

Survey of London 1977 **39** *The Grosvenor Estste in Mayfair* Part 1. London: The Athlone Press

Survey of London 1980 **40** *The Grosvenor Estate in Mayfair* Part II. London: The Athlone Press

Survey of London 1986 **42** *South Kensington: Kensington Square to Earl's Court*. London: The Athlone Press

Tavernor, R and Schofield, R (eds) 1997 *Andrea Palladio: the Four Books of Architecture*. Cambridge, Mass and London: MIT Press

The Ambulator or Stranger's Companion in a Tour Round London 1782. London
Thieme, U and Becke, F 1922 *Allgemeines Lexikon der Bildenden Kunstler* **26**. Frankfurt: Rotten and Loening
Thurley, S 2003 *Hampton Court: A Social and Architectural History*. New Haven and London: Yale University Press
VCH Essex 1973 *The Victoria History of the County of Essex,* **6**. Oxford: Oxford University Press
VCH Middlesex 1976 *The Victoria History of the County of Middlesex,* **5**. Oxford: Oxford University Press
VCH Middlesex 1982 *The Victoria History of the County of Middlesex,* **7**. Oxford: Oxford University Press
VCH Middlesex 1995 *The Victoria History of the County of Middlesex,* **10**. Oxford: Oxford University Press
Vitruvius, P M (trans by Morgan, M H 1914) *The Ten Books of Architecture*. Cambridge, Mass: Harvard University Press
Walpole, H, 1788a *The Gentleman's Magazine*, 1788, 842, 930 and 1070 [obit notices]
Walpole H 1788b *The London Magazine* **44**, 27–30 Sept 1788, 319 [obit notice]
Ware, I 1756 *Complete Body of Architecture Adorned with Plans and Elevations from Original Designs … .* London: T Osborne and J Shipton
Watkin, D 1996 *Sir John Soane, Enlightenment Thought and the Royal Academy Lectures*. Cambridge: Cambridge University Press
White, R 1982 'Danson Park'. *Archaeol Cantiana*, **92** (1982) 51–66
White, R 2003 *Witley Court and Gardens*. London: EH (guidebook)
Wilson, M I 1968 *The English Chamber Organ: History and Development, 1650–1850*. Oxford: Cassirer
Wooller, O 2000, *The Great Estates. Six Country Houses in the London Borough of Bexley*. Bexley: Bexley LSAC
Worsley, G 1991, 'Stiff but not dull'. *Country Life*, **185**, 25 Jul 1991, 90–3
Worsley, G 1995 *Classical Architecture in Britain. The Heroic Age*. New Haven and London: Yale University Press
Worsley, G 2004 *The British Stable*. New Haven and London: Yale University Press
Wright, A 1950 *Joseph Spence, a Critical Biography*. Chicago: Chicago University Press

Unpublished reports

Carr, M C 1970 'The growth and characteristics of a metropolitan suburb, Bexley, N W Kent' (Unpublished PhD thesis, Univ of London)
Fidler, J and Stewart, J 2002 *Danson House, Bexley, Kent* EH BCRT Advisory Report, Sept 2002
Greenhalgh, M 1995 'Gentlemen londoners and the middle classes of Bromley, 1840–1914' (Unpublished PhD thesis, Univ of Greenwich)
Groves, C 2002 *Dendrochronological analysis of conifer timbers from Danson House and Danson Stables, Bexley, Kent* EH Centre for Archaeology Report, 69/2002
Hughes, H 2002 *Paint analysis, Danson House* EH Conservation Studio Report
Hughes, H 2007 *Treatment report, Danson House decorative paintings, Charles Pavillon* EH Conservation Studio Report
Lea, R 2002a *The missing ceiling panel decoration in the saloon at Danson* EH HART Reports and Papers first ser **62** 2002
Lea, R 2002b *Danson, the reconstruction of the service stair balustrade on the landing* EH HART Reports and Papers first ser **63** 2002
Lea, R 2002c *Danson, the reconstruction of the Georgian mirrors in the dining Room, saloon and library* EH HART Reports and Papers **65** 2002
Lea, R 2003a *The saloon paintings by Claude-Joseph Vernet, George Barrett and Elias Martin* EH HART Reports and Papers first ser **64** 2003
Lea, R 2003b *Excavations at Danson: evidence for the service wings, ha-ha and water supply* EH HART Report first ser **80** 2003
Lea, R and Miele, C 2001 *The bookcase and organ scheme of 1766 in the library at Danson* EH HART Reports and Papers first ser **56** 2001
Piper, S, *Danson Hill, Kent* (1974), typescript hand list of leases relating to Danson in Bexley LSAC, 928.8
Potter, G 2003 *The garden of Danson House, Welling, London DA6, London Borough of Bexley, an archaeological evaluation, site code DNN03*. London: Compass Archaeology
Sanderson, R W 1994 *Identification of stone samples from Danson Park* (letter and report, 4 Jun 1994; EH Historic Analysis files, HA 054094/01)

INDEX

Page references in **bold** refer to illustrations.

A

Adam, Robert 15, 16, 44, 55, 61
Adye, John 1
agricultural land 1, 8, **62**, 63, 70
almshouse charity 2, **3**
ancient sculpture, copies 71
ancient stone vase 61, **61**, 71
art collection viii, 52
Asgill House **17**, 18, **19**
attic floor 25, **27**, 60
auctions 52, 66–7, 71, 76
Austin Friars 5

B

'Bailiff's Residence' 67
Barlaston Hall **17**, 19, 20, 23, 34, 39
Barrett, George viii, **viii**, 28, 31, 52, **52**, 67, **73**, 82–3
Bartoli, Pietro Santi 57, **57**
basement floor 33–4, **35**
Bass Taverns Ltd 81
bathroom 74
bays 19, 20, **20**, 23, 26, 63
Bean, Alfred ix–x, 72–5, **72**
Bean, Ann 75
Bean and Jackson 72
bedroom floor 58–9, 59–60, **60**
   Alfred Bean's modernisations **74**, 75
   landing 58, **59**
   local history museum 77, **77**
   masonry and timber structure 23, **24**, 25, **26**

bell-pulls 74
Bettenson, Richard 3
Bettenson, Sir Edward 3
Bexley
   in the 18th century 1
   local studies collection 9, 41, 82
   St Mary the Virgin 68
   Styleman's Almshouses 2, **3**
   Urban District/Borough Council/London Borough 76, 77, 81, 82
Bexley Heritage Trust 81, 82, 83
Bexleyheath, Chapel of Ease/Christ Church 68
Binney, Marcus 15, 16, 18, 19
Blendon 8, **12**
bookcases 55, **56**, **58**, **73**, 74, 82
Boyd, Catherine (née Chapone) (2nd wife of John) 30–1
Boyd, Jean Auguste (Augustus) (father of John) 5–6, 7, 30
Boyd, Jean (grandfather of John) 5
Boyd, John vii, viii–ix
   family background 5–6
   early years 6, 7
   marriage and children 7, 30–1, 61, 63
   takes lease on Danson, 1753 7–8
   acquires freehold of Danson, 1759 8
   acquires freeholds of surrounding properties 8–9, **8**
   rebuilding of Danson 9–13, 30, 31
   commissions paintings and decorative designs 42, 51–5
   engagement of Robert Taylor on other projects, 1760s/70s 64–5
   expansion of estate, 1760s/80s 63

continental tour, 1775 61
financial difficulties, 1780s 61, 63
library collection 61, 67
painting collection viii, 52, 67
Boyd, John II 63, 65–7
   handwritten inventory, 1805 88
Boyd, Mary (née Bumpstead) (1st wife of John) 7, 30, **30**, 58
Boyd and Co 6–7, **6**, 7, 63
Boyd and Pechell 6
Boyd family business ix, 5–7, **5**, 63
Braxton Lodge 15
Brayley, Edward 10
breakfast room 37–8, **38**, 75
Brettingham, Matthew 18
brickwork *see* masonry and timber structure
bridge **29**, 31, **32**
Brown, Lancelot ('Capability') 10
'Building at Risk' x
Bumpstead, Mary 7, 30, **30**
butler's pantry 75

C

Camden Place 66
canal 2, **2**, 3
canted bays 19, 20, **20**, 23, 26, 63
'Capability' Brown 10
ceilings
   entrance hall 40
   saloon 50, **50**, **51**
Chambers, William
   chimney pieces 38, **38**, 43, 44, 46, 54–5, **55**, 57, **57**, 78
   comparisons with Robert Taylor 15, 16

INDEX

design for a Corinthian doorcase 31, **31**
designs for a bridge and a 'temple' 31–2, **32**, 77
main entrance pediment **30**, 31
picture frames 31, 54, **54**, 83
Champneys, Sir John 1
Chantrell, R D 29
Chapel of Ease, Oaklands Road 68
Chapel House *see* Gothick Cottage
Chapone, Catherine 30–1
Chas Collinge & Co 74
chimney pieces
    bedroom floor 60
    breakfast room 37–8, **38**
    dining room 31, **43**, 44, **46**, 48
    library 57, **57**
    loss and recovery 78, 81
    principal bedroom 74
    saloon 54–5, **55**
chimney pots 65
Chineys House 4, **4**, 31
Chinoiserie motifs 51
Chippendale, Thomas 55
Christ Church, Bexleyheath 68
Chute Lodge **17**, 19
circulation space 35, 37, 74–5; *see also* stairs
Cooper Hewitt Museum 51
Coptfold Hall **17**, 19
Corbould, Richard **29**, 63
cornices
    bedroom 59–60
    eaves 22, **22**
    saloon 50, **50**, 51
Coxe, Burrell and Foster 66, 67
Crowther, John **5**

D

damask wallpaper 50–1, 75
Dance, George, the Elder **15**, 16
Dance, George, the Younger 66
Danson (House); *see also* under headings for specific rooms and features
    precedents 16–19, **17**
    design and construction 20–7
    revision of original design 29, 33, 39, 40, 51
    John Boyd II's changes, 1800s 65
    Victorian modernisation and repairs 72
    open to public, 1925 76–7
    wartime use 77
    post-war repairs 77
    listed grade II, 1953 77
    closure to the public 1970 78
    restoration 79–83
    reopening after restoration 83
Danson ('Mannor of Danson') 1–2, **3**, 4
Danson estate
    origins 1–4
    estate plan, 1684 **2**
    lease plan, 1753 **4**
    incorporation of adjacent farms, 1760s/1780s 63
    Revd Joseph Spence's notes on, 1763 85–7
    sale to John Johnston, 1805 66–7, 88
    estate plan, 1805 62
    sale to Alfred Bean, 1863 72
    sale of farmland for development 1922/1923 x, 75
    purchase by Bexley Urban District Council, 1924 76
Danson Farm **62**, 67
Danson Lane 69, **70**
Danson Road 4
de Montfaucon, Bernard 55
decorative schemes 31, 35, 39
    Alfred Bean's modernisations 75
    dining room 40–9, **40**, **41**
    entrance hall 39–40, **39**
    John Boyd II's changes 65
    library 55–7, **56**, **57**, 65
    restoration 80–2
    saloon 50–1, **50**, **51**, **52**, 54–5, **54**, **55**, 82
    stairwell 58–9, **59**
dining room 31, 40–9, **40**, **41**
    Alfred Bean's modernisations **72**, 74, 75
    chimney pieces 31, **43**, 44, **46**, 48
    as evidence of dating of house 28
    masonry and timber structure 23, 39
    proportions 22
    restoration 82
dome, staircase 58, 59, **59**
door furniture 74
door pediments **39**, 39, 82; *see also* entrance door
Doric Temple 68
'double-floor' construction 24, **24**
Du Cane, Peter 15

E

England, George 57
English Heritage x, 42, 79, 81, 82
entrance door 23, **30**, 31, 40, 44, 80, **82**
entrance hall 23, 39–40, **39**, **82**
    butler's pantry/shop 75
    heating arrangements 74
    proportions 22
    restoration 91
entry to the house 29, 37; *see also* main entrance
estate plans **2**, 61, **62**, 63, 66, 68
exterior walls 31, 79, 79–80, **79**, **80**

F

farms viii, **62**, 63, 67, 70
Farrington, Mary 3
Farrington, Thomas 3
fireplaces 60, 74; *see also* chimney pieces
Fisher, Thomas 10, 30
floor frames 25
flooring 24, 39, 75

G

gamekeeper's lodge 69, **70**
garret floor 60

99

Georgian Group Architectural Award 83
Georgian villas **15**, 16–19, **17**, **19**
glazing 40, 59, 72, 74
Gobelins workshops 42
Gothick Cottage **12**, **13**
Gothick-style gamekeeper's lodge 69, **70**
Grafton Street (no 4) 39, (no 7) **64**, 65
grand stair *see* stairwell
grape arbour 67
Great Chapel Field 8–9
Great Pond 4
'Green House' 61
ground floor **33**, 34–5, 37–8, **37**

H
ha-ha 12–13, **13**, 83
hall *see* entrance hall
Hall Place 1
Hamilton, Gavin 61
Harleyford Manor **17**, **18**, 19, 34, 39
Hasted, Edward 10, 29–30
Haworth Hall 57, **57**
heating arrangements 74
Hobbs & Co 74
Home Farm **62**, 67
hothouses 63, 67
Houghton, Sir Henry 65
housekeeper's room 37

I
Ice House 4
interior decoration *see* decorative schemes
internal layout
    Alfred Bean's modernisations 74–5
    basement floor **35**
    bedroom floor 58–9, **60**
    ground floor 34, 37, **37**
    principal floor **38**, 39
inventory, 1805 88

J
Johnston, Hugh 69–70
Johnston, John **61**, 65, 66, 67, 68–9, **68**
Johnston, Sarah **41**, 50, 54, **58**, 70, **71**, 80–1, 82
Jones, Inigo 31

K
kitchen 33–4, **34**, 74–5
Kitchen Offices 28
kitchen wing **ix**, 28, **28**, **33**, 65

L
Ladywell, Lewisham House 7, **7**
lake 9, 31, 32, **32**
land holdings 8, **62**, 63, 70
landscaping *see* park
Langley, Batty 55
layout *see* internal layout
lease plan, 1753 4
Lewisham House 7, **7**
library 31, 55–8, **56**
    Alfred Bean's modernisations **73**, 74, 75
    John Boyd II's changes, 1800s 65
    John Boyd's collection 61
    masonry and timber structure 23, **24**, 25, 39
    proportions 22
    restoration 81, 82
library collection 61, 67
lighting 40, **73**, 74
lime kiln 67
Little Danson **69**, 70, **70**
local history museum 77, **77**
lodges, gate 68–9, **69**, 70
London suburbanisation 1

M
main entrance 29, **30**, 31
Malton, Thomas
    aquatint engraving of Danson viii, **ix**, 14, 28
    engraved plan of the principal floor **ix**, 28
    engraving of Asgill House **19**
'Mannor of Danson' 1–2, **3**
Mansion House 15, **15**
Marlborough House 54–5, **55**
Martin, Elias
    pen-and-ink sketch of the canal and house **3**, 4, 32
    view of the house from the south 31, **52**
masonry and timber structure 23–5, **23**, 24–7
    canted bays **23**, **26**, 63
    restoration 79–80
    trussed partitions **26**, **60**
    Victorian repairs 72
Miller, William **14**
mirrors
    dining room 40, 41, **41**, **42**, 72, 74, 82
    saloon **50**, 74
Morris, Robert 10, 20, **20**
Mount Clare **17**, 18

N
niches 39–40, **39**
Norfolk House 18

O
Oaks, The, Carshalton, Surrey 38, 57, **57**
orangery **66**, 68, **69**
organ 57–8, **58**, 77
    in decorative scheme 55, **56**
    as evidence of dating of house 28
    removal and reinstatement 78, 82
Osterley Park 44
Ottershaw Park 51

P
paintings
    dining room 31, 41–9, **42–9**

John Boyd's collection viii, 52, 67
overdoor and overmantel **frontis**, 42, **43**, **44**, 51–4, **53**, 83–4
saloon 51–2, **52**, **53**
paintwork
    Alfred Bean's modernisations 75
    bedroom floor 59
    dining room 40
    restoration 80
    saloon 50
    stairwell dome 59
Palladio, Andrea vii, 13, 22, **22**, 32
park
    Nathaniel Richmond's plan, c 1762–3 9–10, **9**, 13, 29
    Joseph Spence's measured plan, 1763 4, **11**
    John Johnston's changes 68
    public access 76
    restoration 83
Parker, Matthew 1
partition walls 23, **26**, **60**, 74
Pavillon, Charles 31, 41–9, **42–9**
    order for Vernet's painting 52
    paintings as evidence for dating of house 28
    Sacrifice to Bacchus **frontis**, 42, 43
Pechell, James 6
Peper Harow House 44
perron 11
Peters, Lucy 6
picture frames 41, 54, **54**, 83
Pillement, Jean Baptiste 51
Piranesi Vase 61, **61**, 71
plan form **17**, **18**, 19
plantations 65, 66, **66**, 68
plasterwork
    bedroom floor 59
    dining room 40–1
    entrance hall 39, **39**
    ground floor 35
    library 55
    relief plaques 55, 57, **57**, 65, 82

saloon 50, 51
plumbing 74; *see also* water supply
principal floor **38**, 39–58
    Alfred Bean's modernisations 75
    division of large room, 1800s 65
    masonry and timber structure **25**
proportion and structure 20–7
proportional study **21**, 22
public access 76–7, 78, 83
Purbrook House 17
Purcell Miller Tritton 79

Q
quadrant walls **ix**, 28, **28**, 33

R
radiators 74
relief plaques 55, 57, **57**, 65, 82
Repton, George Stanley 68–9, **69**, 70, **70**
Repton, Humphry 65
restoration 79–83
Richmond, Nathaniel **9**, 10, 13, 29
Rodriguez, Mr 2
Roman sculpture, copies 71
roof structure 25, **27**
room layout *see* internal layout
room proportions 22
roundels 39, **39**, 55, **56**, 81
Royal Institute of British Architects (RIBA) Commendation 83

S
Sacrifice to Bacchus (painting) **frontis**, 42, 43
saloon 50–5, **50**
    Alfred Bean's modernisations 31, **73**, 75
    restoration **83**
sanitation 74
sculpture 39–40, 71
    pediment above main entrance **30**, 31

Piranesi Vase 61, **61**, 71
Selwyn, George Augustus 3
Selwyn, John 2–4
Selwyn, Mary (née Farrington) 3
servants' hall 37
service entrance 37
service rooms 34
service stairs 39, 74
Sharpham House **17**, 19
Skaife, Thomas 35
skylight (dome) 58, 59, **59**
Smith, Captain Nathaniel 63, 66
Soane, Sir John 28, **29**
Somerset House 30
Spence, Revd Joseph 10–12
    notes on the landscape at Danson, 1763 85–7
    sketch plan of proposed new park, 1763 4, **11**, 29
    sketch plan of the water supply to the house 85
St Mary the Virgin, Bexley 68
St Paul's Walden Bury 32, 68, 77
stable block (1802–4) 65–6, **67**, 77–8, **78**, 81, **81**
stable wing **ix**, 28, **28**, 33, 65
stables, preferred location 10
stairs
    main staircase 59, **59**
    service stairs 37, 39, 74
stairwell 58–9
Stevens, Mary 2
stone vase 61, **61**, 71
stone yard 65, 67, 68
stonework 23, **23**, 30, 31, 79–80, **79**, **80**
structural framework *see* masonry and timber structure
Styleman, Francis 2
Styleman, John 2
Styleman, Mary 8
Styleman's Almshouses 2, **3**
suburbanisation 1
Summerson, John 15

INDEX

T

Tapley and Eadworthy 79
Taylor, Laurie 78
Taylor, Michael Angelo 15
Taylor, Robert 14–19, **14**
- architectural career 14–16
- attribution of Danson House design viii, 14
- attribution of Gothick Cottage 12
- Danson House commission 13, 14
- designs for Danson House 18, 19, 20, 22, 30, 31; *see also* headings for specific rooms and features
- engravings of his work viii, **16, 19, ix**
- other work 7, 15–16, 18–19, 64–5
- and villa design 13, 16

tea room 76–7, **77**
temple
- as shown on 1753 lease plan 4
- Chamber's Doric 31–2, **32**, 68, 77

Thauvet, Andrew 5, 6
Thorpe, Charles **12**
timber structure *see* masonry and timber structure
Trevanion, John 7, 66
trussed partitions **26, 60**

U

Upper Brook Street (no 33) 64, **64**

V

vase, ancient stone 61, **61**, 71
Vernet, Claude-Joseph 31, 51–2, **53**, 54, **54**, 83
villa designs **15**, 16–19, **17, 19**
Vitruvius 19, 22

W

wall coverings 50–1, **51**, 75, 82
wall plaques 55, 57, **57**, 65, 82

walls
- house exterior 31, 79–80
- quadrant **ix**, 28, **28**, 33

Ware, Isaac 10
wartime use 77
water supply **33**, 34, **85**
Wilton, Joseph 44
Wilton House 55
window shutters 60, 74
wings **ix**, 28, **28, 33**, 65
'Winter Garden' 68
woodwork 23; *see also* bookcases
Wright, Thomas 4

Y

Yenn, John 54, **54**

Z

Zuccarelli, Francesco 42